Uniqueness and Stability in Determining a Rigid Inclusion in an Elastic Body

MEMOIRS
of the
American Mathematical Society

Number 938

Uniqueness and Stability in Determining a Rigid Inclusion in an Elastic Body

Antonino Morassi
Edi Rosset

July 2009 • Volume 200 • Number 938 (third of 6 numbers) • ISSN 0065-9266

American Mathematical Society
Providence, Rhode Island

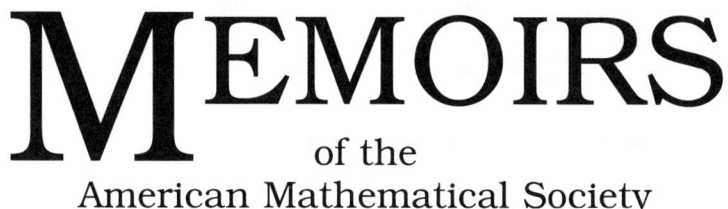

2000 *Mathematics Subject Classification.*
Primary 35R30; Secondary 35R25, 35J55, 74B05.

Library of Congress Cataloging-in-Publication Data

Morassi, Antonino
Uniqueness and stability in determining a rigid inclusion in an elastic body / Antonino Morassi and Edi Rosset.
p. cm. — (Memoirs of the American Mathematical Society, ISSN 0065-9266 ; no. 938)
"Volume 200, number 938 (third of 6 numbers)."
ISBN 978-0-8218-4325-3 (alk. paper)
1. Inverse problems (Differential equations)—Numerical solutions. 2. Numerical analysis—Improperly posed problems. 3. Elasticity—Mathematical models. I. Rosset, Edi, 1961– II. Title.

QA377.M667 2009
518′.64—dc22 2009008260

Memoirs of the American Mathematical Society

This journal is devoted entirely to research in pure and applied mathematics.

Subscription information. The 2009 subscription begins with volume 197 and consists of six mailings, each containing one or more numbers. Subscription prices for 2009 are US$709 list, US$567 institutional member. A late charge of 10% of the subscription price will be imposed on orders received from nonmembers after January 1 of the subscription year. Subscribers outside the United States and India must pay a postage surcharge of US$65; subscribers in India must pay a postage surcharge of US$95. Expedited delivery to destinations in North America US$57; elsewhere US$160. Each number may be ordered separately; *please specify number* when ordering an individual number. For prices and titles of recently released numbers, see the New Publications sections of the *Notices of the American Mathematical Society.*

Back number information. For back issues see the *AMS Catalog of Publications.*

Subscriptions and orders should be addressed to the American Mathematical Society, P. O. Box 845904, Boston, MA 02284-5904, USA. *All orders must be accompanied by payment.* Other correspondence should be addressed to 201 Charles Street, Providence, RI 02904-2294, USA.

Memoirs of the American Mathematical Society (ISSN 0065-9266) is published bimonthly (each volume consisting usually of more than one number) by the American Mathematical Society at 201 Charles Street, Providence, RI 02904-2294, USA. Periodicals postage paid at Providence, RI. Postmaster: Send address changes to Memoirs, American Mathematical Society, 201 Charles Street, Providence, RI 02904-2294, USA.

This publication is indexed in *Science Citation Index®*, *SciSearch®*, *Research Alert®*, *CompuMath Citation Index®*, *Current Contents®/Physical, Chemical & Earth Sciences.*
Printed in the United States of America.

♾ The paper used in this book is acid-free and falls within the guidelines
established to ensure permanence and durability.
Visit the AMS home page at `http://www.ams.org/`

10 9 8 7 6 5 4 3 2 1 14 13 12 11 10 09

Contents

Abstract

We consider the inverse problem of determining a rigid inclusion inside an isotropic elastic body Ω, from a single measurement of traction and displacement taken on the boundary of Ω. For this severely ill–posed problem we prove uniqueness and a conditional stability estimate of log–log type.

Received by the editor June 15, 2005.

2000 *Mathematics Subject Classification.* Primary 35R30; Secondary 35R25, 35J55, 74B05.

Key words and phrases. Inverse problems, linearized elasticity, rigid inclusion, uniqueness, stability estimates, unique continuation.

The first author was supported in part by MIUR, PRIN # 2003082352.

The second author was supported in part by MIUR, PRIN # 2004011204.

Acknowledgments

The authors wish to thank Giovanni Alessandrini, Dusan Repovs and Sergio Vessella for stimulating discussions and helpful suggestions.

CHAPTER 1

Introduction

In this paper we consider the inverse problem of identifying a rigid inclusion inside an elastic body Ω from measurements of traction and displacement taken on the boundary. As an example of an application of practical interest, this kind of problems arises in non–destructive testing for damage assessment of mechanical specimens, which are possible defective due to the presence of interior rigid inclusions induced during the manufacturing process. More precisely, let the elastic body Ω be represented by a bounded domain in $\mathbb{R}^2$, or $\mathbb{R}^3$, inside which a possible unknown rigid inclusion D is present. Our aim is to identify D by applying a traction field φ at the boundary $\partial\Omega$ and by measuring the induced displacement field on a portion $\Sigma \subset \partial\Omega$.

Working within the framework of the linearized elasticity, where $\mathbb{C}$ denotes the known elasticity tensor of the material, the displacement field u satisfies the following boundary value problem

$$\begin{cases} \operatorname{div}(\mathbb{C}\nabla u) = 0, & \text{in } \Omega \setminus \overline{D}, \\ (\mathbb{C}\nabla u)\nu = \varphi, & \text{on } \partial\Omega, \\ u_{|\partial D} \in \mathcal{R}, \end{cases} \tag{1.1}$$

coupled with the *equilibrium condition*

$$\int_{\partial D} (\mathbb{C}\nabla u)\nu \cdot r = 0, \quad \text{for every } r \in \mathcal{R}, \tag{1.2}$$

where $\mathcal{R}$ denotes the linear space of the infinitesimal rigid displacements $r(x) = c + Wx$, where c is any constant n–vector and W is any constant skew $n \times n$ matrix. We shall assume $\mathbb{C}$ strongly convex and of Lamé type, satisfying some regularity conditions (see (2.25)). Given any $\varphi \in H^{-\frac{1}{2}}(\partial\Omega, \mathbb{R}^n)$, such that $\int_{\partial\Omega} \varphi \cdot r = 0$ for every $r \in \mathcal{R}$, problem (1.1)-(1.2) admits a solution $u \in H^1(\Omega \setminus \overline{D})$, which is unique up to an infinitesimal rigid displacement. In order to specify a unique solution, we shall assume in the sequel the following normalization condition

$$u = 0 \quad \text{on } \partial D. \tag{1.3}$$

Therefore, the inverse problem consists in determining the unknown rigid inclusion D, appearing in problem (1.1)–(1.3), from a single pair of Cauchy data $\{u, (\mathbb{C}\nabla u)\nu\}$ on $\partial\Omega$.

The indeterminacy of the displacement field u and the consequent arbitrariness of the normalization (1.3) which we have chosen, lead to the following formulation of the uniqueness issue.

Given two solutions u_i to (1.1)–(1.3) *when $D = D_i$, $i = 1, 2$, satisfying*

$$(\mathbb{C}\nabla u_i)\nu = \varphi, \quad \text{on } \partial\Omega, \tag{1.4}$$

$$u_1 - u_{2|_\Sigma} \in \mathcal{R}, \tag{1.5}$$

does $D_1 = D_2$ hold?

Here we prove uniqueness under the assumption that ∂D is of C^1 class, see Theorem 2.3. The main tools which we have employed to prove this uniqueness result are the weak unique continuation principle for solutions to the Lamé system (first established by Weck [**45**]), the uniqueness for the corresponding Cauchy problem, (see, for instance, [**24**] and [**36**]), and geometrical arguments related to the structure of the linear space $\mathcal{R}$ which involve different techniques according to the space dimension (see Lemma 3.1 for the 3–D setting).

From the point of view of stability, it is almost evident that this inverse problem is severely ill–posed. In fact, in order to determine the unknown rigid inclusion D, it seems necessary to estimate the solution u from the Cauchy data on the exterior boundary up to ∂D. Therefore, due to the ill–posedness of the Cauchy problem for elliptic systems, one can expect only a weak rate of continuity, under some a priori information on the unknown boundary ∂D.

The stability issue can be formulated as follows.

Given two solutions u_i to (1.1)–(1.3) *when $D = D_i$, $i = 1, 2$, satisfying*

$$(\mathbb{C}\nabla u_i)\nu = \varphi, \quad \text{on } \partial\Omega, \tag{1.6}$$

$$\min_{r\in\mathcal{R}} \|(u_1 - u_2) - r\|_{L^2(\Sigma)} < \epsilon, \quad \textit{for some } \epsilon > 0, \tag{1.7}$$

to evaluate the rate at which the Hausdorff distance between D_1 and D_2 tends to zero as ϵ tends to zero.

In the present paper, assuming $C^{1,\alpha}$ regularity of ∂D, $0 < \alpha \leq 1$, we prove a constructive stability estimate of log–log type under suitable a priori assumptions, see Theorem 2.5 for a precise statement.

Our approach to prove stability is essentially based on quantitative estimates of unique continuation, precisely: a three spheres inequality for solutions to the Lamé system, which was obtained in [**5**] (see Proposition 6.5); stability estimates for solutions to the Cauchy problem, obtained in [**36**] (see Proposition 6.4); a stability estimate of continuation from the interior for a mixed problem (see Proposition 4.1). The analogous version of this last estimate for the Neumann problem was obtained in [**36**] (see Proposition 5.1); here, in order to treat the mixed problem, it has been crucial to derive a constructive version of a Korn–type inequality for maps vanishing on a portion of the boundary (see Proposition (5.3)). Moreover, in the stability context, due to the general form of the condition (1.7), the complications of geometrical character arising in the proof of uniqueness become significantly harder. To overcome these difficulties, we have derived a geometrical result (see Lemma 7.1), which turns out to be a crucial ingredient to prove stability in the three dimensional case.

The inverse problem considered here can be cast into the (by now) wide field of inverse boundary value problems. The prototypical model, and perhaps the most well–known, is the so–called *inverse conductivity problem* of Calderón [**15**], also known as the problem of electrical impedance tomography.

This problem deals with the determination of the scalar coefficient $\gamma = \gamma(x)$ modeling the electrical conductivity of an electrically conducting body Ω from measurements of voltage and current taken at the boundary. In mathematical terms, one is dealing with a scalar elliptic equation in divergence form and the uniqueness and the stability issues correspond to injectivity and continuity of the inverse for the map

$$\gamma \mapsto \Lambda_\gamma, \tag{1.8}$$

where Λ_γ is the so–called *Dirichlet-to-Neumann* map

$$\Lambda_\gamma : H^{\frac{1}{2}}(\partial\Omega) \to H^{-\frac{1}{2}}(\partial\Omega) \tag{1.9}$$
$$g \mapsto \gamma\nabla u \cdot \nu,$$

$u \in H^1(\Omega)$ being the solution to the boundary value problem

$$\operatorname{div}(\gamma\nabla u) = 0, \quad \text{in } \Omega, \tag{1.10}$$

$$u = g, \quad \text{on } \partial\Omega. \tag{1.11}$$

Regarding the uniqueness issue, the cornerstones of the theory are due to Kohn and Vogelius [**30**], [**31**], Sylvester and Uhlmann [**43**] for the case of space dimension $n \geq 3$, Nachman [**37**] for the case of dimension $n = 2$. The stability issue has been treated by Alessandrini [**3**] for $n \geq 3$, and by Liu [**35**] and Barceló, Barceló and Ruiz [**11**] for $n = 2$. See, for further details and references, Isakov [**29**], Borcea [**13**] and Uhlmann [**44**].

A corresponding theory has been developed for the analogous problem for linearized elasticity and regarding uniqueness we can refer to Nakamura and Uhlmann [**39**], [**40**], [**41**], [**42**] and Eskin and Ralston [**25**], see also the review paper by Nakamura [**38**]. Unfortunately, this theory is not yet as complete as for the scalar case.

The above described theories require the complete knowledge of all Dirichlet and Neumann data at the boundary, and, in most cases, a high degree of smoothness is a-priori assumed on the unknown parameters (either the conductivity γ in the scalar case, or the Lamé parameters μ, λ in the elasticity model). In practice, however, only finitely many boundary measurements can be collected, and the unknown parameters may be discontinuous or even attain to extreme values where the boundedness or the ellipticity constraints are violated. In the scalar case of the conductivity model, this is the case when perfectly conducting ($\gamma = \infty$) or perfectly insulating ($\gamma = 0$) inclusions are present.

For this reason, one line of current research of this field is the one of determining unknown regions (inclusions) where the parameters attain to extreme values from the knowledge of a single boundary measurement. That is, one pair of Dirichlet and Neumann data. For this class of inverse problems the role of the unknown is played by the inclusion or, equivalently, by its boundary, which is parametrized by $(n-1)$ independent variables, so that one is mainly dealing with an inverse problem of geometrical character for which it is expected that only one boundary measurement, which is represented as a function of $(n-1)$ variables, is sufficient to recover the inclusion. Regarding the uniqueness and stability results for such class of problems we refer to Beretta and Vessella [**12**], Alessandrini and Rondi [**8**], Alessandrini, Beretta, Rosset and Vessella [**4**], Bukhgeim, Cheng and Yamamoto [**14**] and Cheng, Hon and Yamamoto [**19**].

Typically the stability results are of logarithmic type, and examples constructed by Alessandrini and Rondi [**8**] and by Di Cristo and Rondi [**22**], [**23**] show that indeed such rate of stability is the best possible.

The corresponding inverse problems in the field of elasticity deal with determination of void subsets (cavities) of the elastic body or else of rigid subsets immersed in the elastic material. The boundary measurements in this case are given by one pair of displacement (Dirichlet data) and traction field (Neumann data) on the boundary. The former problem (the one of determining cavities) has been treated in [**36**], see also [**10**] for a uniqueness result. The latter (of determination of rigid inclusions) is the object of the present note.

It is worth examining in more detail the comparison of this problem with the related problems in the scalar case. In fact, in the conductivity model, a perfectly conducting inclusion is an inclusion where the electrostatic potential is *constant*, but the value of the constant is not a–priori known. The problem has been treated, in the 2–D setting, in [**8**].

Another relevant problem arises when one considers the scalar equation (1.10) as a model of the stationary thermal conduction. In [**4**] the authors faced the problem of the detection of a solidification surface inside a thermic conducting body from a single measurement of temperature and heat flux at the boundary. Since the value of the melting temperature is a known parameter, the Dirichlet boundary condition on the unknown surface is, in this case, a known constant value. Therefore the complications due to the indeterminacy of the solution above described for the elasticity context do not occur in the thermal setting. In addition, for scalar elliptic equations in divergence form a further basic tool, called *doubling inequality at the boundary* is available (see [**2**], [**33**], [**1**]). This inequality represents a quantitative form of the unique continuation principle at the boundary and turns out to be the key ingredient to obtain a better modulus of continuity, precisely a log–stability estimate under $C^{1,1}$ regularity assumptions on the unknown surface (see [**4**, Theorem 2.2]).

In Chapter 8, as a byproduct of our present approach, we fill this gap in the scalar theory and prove optimal stability estimates of logarithmic type for the scalar inverse problem of determining a perfectly conducting inclusion in any space dimension.

Let us now comment on the main differences with the problem of determining cavities inside an elastic body and the results obtained in [**36**]. In this case one has homogeneous boundary conditions of Neumann type on ∂D and, again, the solution u of the corresponding direct problem is determined up to an infinitesimal rigid displacement.

In [**36**] we have obtained a log–log stability estimate under the (apparently) stronger assumption

$$\|u_1 - u_2\|_{L^2(\Sigma)} < \epsilon, \quad \text{for some } \epsilon > 0, \tag{1.12}$$

instead of (1.7). Indeed, in this case, the choice (1.12) turns out to be not restrictive since the homogeneous Neumann boundary condition on ∂D remains valid also for $u_1 - r$ and $u_2 - r$, for any $r \in \mathcal{R}$. This is not the case, of course, when, as in the situation considered in the present paper, homogeneous Dirichlet conditions on ∂D are assumed.

The plan of the paper is as follows. In Chapter 2 we state the main results, Theorem 2.3, Theorem 2.5 and Corollary 2.7. In Chapter 3 we prove the uniqueness result. In Chapter 4 we state the auxiliary propositions concerning the estimates of continuation from the interior (Proposition 4.1), and from Cauchy data (Propositions 4.2 and 4.3), and we give the proof of Theorem 2.5. In Chapter 5 we state and prove the Korn–type inequality suitable for our purposes and we prove Proposition 4.1. Chapter 6 contains the proofs of Proposition 4.2 and Proposition 4.3, concerning stability estimates of continuation from Cauchy data in the two–dimensional case, whereas the three–dimensional case is treated in Chapter 7. Finally, in Chapter 8 we discuss the related inverse problem of determining a perfectly conducting inclusion arising in the electrostatic framework.

CHAPTER 2

Main results

2.1. Notation and definitions

When representing locally a boundary as a graph, we shall use the following notation. For every $x \in \mathbb{R}^n$ we set $x = (x', x_n)$, where $x' = (x'_1, ..., x'_{n-1}) \in \mathbb{R}^{n-1}$, $x_n \in \mathbb{R}$, $n = 2$ or $n = 3$.

DEFINITION 2.1. ($C^{k,\alpha}$ regularity) Let Ω be a bounded domain in $\mathbb{R}^n$. Given k, α, with $k \in \mathbb{N}$, $0 < \alpha \leq 1$, we say that a portion S of $\partial\Omega$ is of *class* $C^{k,\alpha}$ *with constants* ρ_0, $M_0 > 0$, if, for any $P \in S$, there exists a rigid transformation of coordinates under which we have $P = 0$ and

$$\Omega \cap B_{\rho_0}(0) = \{x \in B_{\rho_0}(0) \quad | \quad x_n > \psi(x')\},$$

where ψ is a $C^{k,\alpha}$ function on $B'_{\rho_0}(0) = B_{\rho_0}(0) \cap \{x_n = 0\} \subset \mathbb{R}^{n-1}$ satisfying

$$\psi(0) = 0,$$

$$\nabla\psi(0) = 0, \quad \text{when } k \geq 1,$$

$$\|\psi\|_{C^{k,\alpha}(B'_{\rho_0}(0))} \leq M_0\rho_0.$$

When $k = 0$, $\alpha = 1$, we also say that S is of *Lipschitz class with constants* ρ_0, M_0.

REMARK 2.2. We use the convention to normalize all norms in such a way that their terms are dimensionally homogeneous and coincide with the standard definition when the dimensional parameter equals one. For instance, the norm appearing above is meant as follows

$$\|\psi\|_{C^{k,\alpha}(B'_{\rho_0}(0))} = \sum_{i=0}^{k} \rho_0^i \|D^i\psi\|_{L^\infty(B'_{\rho_0}(0))} + \rho_0^{k+\alpha}|D^k\psi|_{\alpha,B'_{\rho_0}(0)},$$

where

$$|D^k\psi|_{\alpha,B'_{\rho_0}(0)} = \sup_{\substack{x',\, y' \in B'_{\rho_0}(0) \\ x' \neq y'}} \frac{|D^k\psi(x') - D^k\psi(y')|}{|x' - y'|^\alpha}.$$

Similarly, we set

$$\|u\|_{H^1(\Omega,\mathbb{R}^n)} = \left(\int_\Omega u^2 + \rho_0^2 \int_\Omega |\nabla u|^2\right)^{\frac{1}{2}},$$

and so on for boundary and trace norms such as $\|\cdot\|_{H^{\frac{1}{2}}(\partial\Omega,\mathbb{R}^n)}$, $\|\cdot\|_{H^{-\frac{1}{2}}(\partial\Omega,\mathbb{R}^n)}$, $\|\cdot\|_{H^{-1}(\partial\Omega,\mathbb{R}^n)}$.

For any $h > 0$ we denote

$$\Omega_h = \{x \in \Omega \mid \mathrm{dist}(x, \partial\Omega) > h\}. \tag{2.1}$$

We denote by $\mathbb{M}^{m\times n}$ the space of $m \times n$ real valued matrices and by $\mathcal{L}(X, Y)$ the space of bounded linear operators between Banach spaces X and Y. When $m = n$, we shall also denote $\mathbb{M}^n = \mathbb{M}^{n\times n}$.

For every pair of real n–vectors a and b, we denote by $a \otimes b$ the $n \times n$ matrix with entries

$$(a \otimes b)_{ij} = a_i b_j, \quad i, j = 1, ..., n. \tag{2.2}$$

For every $n \times n$ matrices A, B and for every $\mathbb{C} \in \mathcal{L}(\mathbb{M}^n, \mathbb{M}^n)$, we use the following notation:

$$(\mathbb{C}A)_{ij} = \sum_{k,l=1}^{n} C_{ijkl} A_{kl}, \tag{2.3}$$

$$A \cdot B = \sum_{i,j=1}^{n} A_{ij} B_{ij}, \tag{2.4}$$

$$|A| = (A \cdot A)^{\frac{1}{2}}, \tag{2.5}$$

$$\widehat{A} = \frac{1}{2}(A + A^T), \tag{2.6}$$

for every $n \times n$ matrices A, B.

Let us introduce the linear space of the infinitesimal rigid displacements

$$\mathcal{R} = \left\{r(x) = c + Wx,\ c \in \mathbb{R}^n,\ W \in \mathbb{M}^n,\ W + W^T = 0\right\}, \tag{2.7}$$

where x is the vector position of a generic point in $\mathbb{R}^n$. Denoting by $\{e_1, e_2, e_3\}$ the canonical orthonormal basis of $\mathbb{R}^3$, by Korn and Poincaré inequalities, it is easy to verify that

$$\mathcal{R} = \{r(x) = c + a \times x\}, \tag{2.8}$$

where $\times$ is the usual vector product in $\mathbb{R}^3$ and

$$c = \sum_{i=1}^{2} c_i e_i, \quad a = a_3 e_3, \quad c_i, a_3 \in \mathbb{R}, \qquad \text{for } n = 2, \tag{2.9}$$

$$c = \sum_{i=1}^{3} c_i e_i, \quad a = \sum_{i=1}^{3} a_i e_i, \quad c_i, a_i \in \mathbb{R}, \qquad \text{for } n = 3. \tag{2.10}$$

Let us notice that, by Korn and Poincaré inequalities, we have

$$\mathcal{R} = \left\{v \in H^1(\mathbb{R}^n, \mathbb{R}^n) \mid \widehat{\nabla} v \equiv 0\right\}. \tag{2.11}$$

2.2. A priori information

i) A priori information on the domain.

We shall assume that Ω is a bounded domain in $\mathbb{R}^n$ such that, given ρ_0, $M_1 > 0$,

$$|\Omega| \leq M_1 \rho_0^n, \tag{2.12}$$

where $|\Omega|$ denotes the Lebesgue measure of Ω.

For the sake of simplicity, we shall assume in the sequel that

$$\partial\Omega \text{ is connected,} \tag{2.13}$$

but we emphasize that all the results stated in the next Chapter continue to hold in the general case when (2.13) is removed, see Remark 2.6 and Remark 6.6.

We shall assume that Ω contains an open connected rigid inclusion D such that

$$\Omega \setminus \overline{D} \text{ is connected,} \tag{2.14}$$

$$\partial D \text{ is connected,} \tag{2.15}$$

and

$$\operatorname{dist}(D, \partial\Omega) \geq \rho_0. \tag{2.16}$$

Moreover, we assume that we can select an open portion Σ within $\partial\Omega$ (representing the portion of the boundary where measurements are taken) such that for some $P_0 \in \Sigma$

$$\partial\Omega \cap B_{\rho_0}(P_0) \subset \Sigma. \tag{2.17}$$

Regarding the regularity of the boundaries, given α, M_0, $0 < \alpha \leq 1$, $M_0 > 0$, we assume that

$$\partial\Omega \text{ is of } \textit{class } C^{1,\alpha} \textit{ with constants } \rho_0, M_0, \tag{2.18}$$

$$\partial D \text{ is of } \textit{class } C^{1,\alpha} \textit{ with constants } \rho_0, M_0, \tag{2.19}$$

and, moreover, that

$$\Sigma \text{ is of } \textit{class } C^{2,\alpha} \textit{ with constants } \rho_0, M_0. \tag{2.20}$$

ii) Assumptions about the boundary data.

On the Neumann data φ appearing in problem (1.1) we assume that

$$\varphi \in H^{-\frac{1}{2}}(\partial\Omega, \mathbb{R}^n), \quad \varphi \not\equiv 0, \tag{2.21}$$

the (obvious) compatibility condition

$$\int_{\partial\Omega} \varphi \cdot r = 0, \quad \text{for every } r \in \mathcal{R}, \tag{2.22}$$

and that, for a given constant $F > 0$,

$$\frac{\|\varphi\|_{H^{-\frac{1}{2}}(\partial\Omega,\mathbb{R}^n)}}{\|\varphi\|_{H^{-1}(\partial\Omega,\mathbb{R}^n)}} \leq F. \tag{2.23}$$

iii) Assumptions about the elasticity tensor.

We assume that the elastic material is *isotropic*, that is the elasticity tensor field $\mathbb{C} = \mathbb{C}(x) \in \mathcal{L}(\mathbb{M}^n, \mathbb{M}^n)$ has components C_{ijkl} given by

$$C_{ijkl}(x) = \lambda(x)\delta_{ij}\delta_{kl} + \mu(x)(\delta_{ki}\delta_{lj} + \delta_{li}\delta_{kj}), \quad \text{for every } x \in \overline{\Omega}, \tag{2.24}$$

where $\lambda = \lambda(x)$ and $\mu = \mu(x)$ are the *Lamé moduli*, see [**28**]. Moreover we assume that the *Lamé moduli* satisfy the $C^{1,1}$ regularity condition

$$\|\mu\|_{C^{1,1}(\overline{\Omega})} + \|\lambda\|_{C^{1,1}(\overline{\Omega})} \leq M. \tag{2.25}$$

and the *strong convexity* condition

$$\mu(x) \geq \alpha_0, \quad 2\mu(x) + n\lambda(x) \geq \beta_0, \quad \text{for every } x \in \overline{\Omega}. \tag{2.26}$$

where M, α_0, β_0 are given positive constants.

Notice that (2.24) implies the following symmetry conditions

$$\mathbb{C}A = \mathbb{C}\widehat{A}, \tag{2.27}$$

$$\mathbb{C}A \quad \text{is } \textit{symmetric}, \tag{2.28}$$

$$\mathbb{C}A \cdot B = \mathbb{C}B \cdot A, \tag{2.29}$$

for every $n \times n$ matrices A, B.

Moreover, (2.26) implies that

$$\mathbb{C}(x)A \cdot A \geq \xi_0 |A|^2, \quad \text{for every } x \in \overline{\Omega}, \tag{2.30}$$

for any symmetric matrix A, with ξ_0 a positive constant only depending on α_0, β_0.

Denoting by I_n the $n \times n$ identity matrix, we have

$$\mathbb{C}(x)A = \lambda(x)(A \cdot I_n)I_n + 2\mu(x)\widehat{A}, \tag{2.31}$$

and the displacement equation of equilibrium becomes the Lamé system

$$\operatorname{div}(2\mu\widehat{\nabla} u) + \nabla(\lambda \operatorname{div} u) = 0, \quad \text{in } \Omega. \tag{2.32}$$

We shall refer to the set of constants α, M_0, M_1, F, α_0, β_0, M as to the *a priori data.*

In the sequel we shall consider the following boundary value problem of mixed type

$$\begin{cases} \operatorname{div}(\mathbb{C}\nabla u) = 0, & \text{in } \Omega \setminus \overline{D}, \\ (\mathbb{C}\nabla u)\nu = \varphi, & \text{on } \partial\Omega, \\ u = 0, & \text{on } \partial D, \end{cases} \tag{2.33}$$

coupled with the *equilibrium condition*

$$\int_{\partial D} (\mathbb{C}\nabla u)\nu \cdot r = 0, \quad \text{for every } r \in \mathcal{R}. \tag{2.34}$$

By standard variational arguments (see, for instance, [**16**]), it is easy to see that problem (2.33)–(2.34) admits a unique solution $u \in H^1(\Omega \setminus \overline{D}, \mathbb{R}^n)$ such that

$$\|u\|_{H^1(\Omega\setminus\overline{D},\mathbb{R}^n)} \leq C\rho_0^{\frac{3}{2}} \|\varphi\|_{H^{-\frac{1}{2}}(\partial\Omega,\mathbb{R}^n)}, \tag{2.35}$$

where $C > 0$ only depends on α_0, β_0, M_0 and M_1.

2.3. Statement of the main results

THEOREM 2.3 (Uniqueness). *Let Ω be a bounded domain satisfying* (2.13) *and having Lipschitz boundary. Let D_i, $i = 1, 2$, be two domains compactly contained in Ω, having C^1 boundary and satisfying* (2.14) *and* (2.15). *Moreover, let Σ be an open portion of $\partial\Omega$ of class $C^{2,\alpha}$. Let $u_i \in H^1(\Omega \setminus \overline{D_i}, \mathbb{R}^n)$ be the solution to* (2.33), (2.34), *when $D = D_i$, $i = 1, 2$, let* (2.21), (2.22) *be satisfied and let the elasticity tensor $\mathbb{C}$ of Lamé type, with Lamé moduli λ and μ of $C^{1,1}$ class satisfying $\mu > 0$, $2\mu + n\lambda > 0$ in $\overline{\Omega}$. If we have*

$$(u_1 - u_2)_{|_\Sigma} \in \mathcal{R} \tag{2.36}$$

then

$$D_1 = D_2. \tag{2.37}$$

REMARK 2.4. Let us briefly comment about a possible weakening of the assumption concerning the regularity of the boundary of the inclusion. In the proof of Theorem 2.3, this regularity assumption is related to the need of continuity of the solution u to (2.33), (2.34) up to the boundary of the inclusion.

In the two dimensional case, the continuity up to the boundary is guaranteed in domains having Lipschitz boundary and it follows, for instance, by adapting arguments by Campanato [**17**], [**18**] and by Giaquinta and Modica [**27**].

Instead, in the three dimensional case, at our knowledge, continuity up to the boundary in Lipschitz domains has been obtained only for elliptic systems with constant coefficients, see [**21**], [**20**] and [**26**].

THEOREM 2.5 (Stability). *Let Ω be a domain satisfying* (2.12), (2.13) *and* (2.18). *Let D_i, $i = 1, 2$, be two connected open subsets of Ω satisfying* (2.14), (2.15) (2.16) *and* (2.19). *Moreover, let Σ be an open portion of $\partial\Omega$ satisfying* (2.17) *and* (2.20). *Let $u_i \in H^1(\Omega \setminus \overline{D_i}, \mathbb{R}^n)$ be the solution to* (2.33), (2.34), *when $D = D_i$, $i = 1, 2$, and let* (2.21)–(2.26) *be satisfied. If, given $\epsilon > 0$, we have*

$$\|u_1 - u_2 - \overline{r}\|_{L^2(\Sigma,\mathbb{R}^n)} = \min_{r \in \mathcal{R}} \|u_1 - u_2 - r\|_{L^2(\Sigma,\mathbb{R}^n)} \leq \rho_0^{\frac{n-1}{2}} \epsilon, \tag{2.38}$$

then we have

$$d_{\mathcal{H}}(\partial D_1, \partial D_2) \leq \rho_0 \omega \left(\frac{\epsilon}{\rho_0^{\frac{3-n}{2}} \|\varphi\|_{H^{-\frac{1}{2}}(\partial\Omega,\mathbb{R}^n)}} \right) \tag{2.39}$$

and

$$d_{\mathcal{H}}(\overline{D_1}, \overline{D_2}) \leq \rho_0 \omega \left(\frac{\epsilon}{\rho_0^{\frac{3-n}{2}} \|\varphi\|_{H^{-\frac{1}{2}}(\partial\Omega,\mathbb{R}^n)}} \right), \tag{2.40}$$

where ω is an increasing continuous function on $[0, \infty)$ which satisfies

$$\omega(t) \leq C(\log|\log t|)^{-\eta}, \quad \textit{for every } t,\ 0 < t < e^{-1}, \tag{2.41}$$

and C, η, $C > 0$, $0 < \eta \leq 1$, are constants only depending on the a priori data.

Here $d_{\mathcal{H}}$ denotes the Hausdorff distance between bounded closed sets of $\mathbb{R}^n$.

REMARK 2.6. Let us notice that for the sake of simplicity we have assumed (2.13), but, in fact, the above theorem holds true in the more general case in which $\partial\Omega$ is not connected and the traction field φ has support contained in the exterior boundary $\partial\Omega^e$ of Ω, defined as the boundary of the unbounded connected component of $\mathbb{R}^n \setminus \Omega$. See also Remark 6.6 for more details.

COROLLARY 2.7. *In the hypotheses of Theorem 2.5, there exist* $\tilde{\rho}_0$, $0 < \tilde{\rho}_0 \leq \rho_0$, *only depending on* ρ_0, M_0, α, *and* $\epsilon_0 > 0$, *only depending on the a priori data, such that if* $\epsilon \leq \epsilon_0$ *then for every* $P \in \partial D_1 \cup \partial D_2$ *there exists a rigid transformation of coordinates under which* $P = 0$ *and*

$$D_i \cap B_{\tilde{\rho}_0}(0) = \{x \in B_{\tilde{\rho}_0}(0) \text{ s.t. } x_n > \psi_i(x')\}, \quad i = 1, 2, \tag{2.42}$$

where ψ_1, ψ_2 *are* $C^{1,\alpha}$ *functions on* $B_{\tilde{\rho}_0}(0) \subset \mathbb{R}^{n-1}$ *which satisfy, for every* β, $0 < \beta < \alpha$,

$$\|\psi_1 - \psi_2\|_{C^{1,\beta}\left(B_{\tilde{\rho}_0}(0)\right)} \leq \rho_0 K \omega(\tilde{\epsilon})^{\frac{\alpha-\beta}{1+\alpha}}, \tag{2.43}$$

where

$$\tilde{\epsilon} = \frac{\epsilon}{\rho_0^{\frac{3-n}{2}} \|\varphi\|_{H^{-\frac{1}{2}}(\partial\Omega,\mathbb{R}^n)}},$$

ω *is as in* (2.41) *and* $K > 0$ *only depends on* M_0, α *and* β. *Furthermore, there exists a* $C^{1,\alpha}$ *diffeomorphism* $F : \mathbb{R}^n \to \mathbb{R}^n$ *such that* $F(D_2) = D_1$ *and for every* β, $0 < \beta < \alpha$,

$$\|F - Id\|_{C^{1,\beta}(\mathbb{R}^n)} \leq \rho_0 K \omega(\tilde{\epsilon})^{\frac{\alpha-\beta}{1+\alpha}}, \tag{2.44}$$

with K, ω *as above. Here* $Id : \mathbb{R}^n \to \mathbb{R}^n$ *denotes the identity mapping.*

CHAPTER 3

Proof of the uniqueness result

PROOF OF THEOREM 2.3. Let G be the connected component of $\Omega \setminus (\overline{D_1 \cup D_2})$ such that $\partial G \supset \Sigma$.

Let $\overline{r} \in \mathcal{R}$ be such that $u_1 - u_2 = \overline{r}$ on Σ and let us denote $w = u_1 - u_2 - \overline{r}$. We have that w satisfies the following Cauchy problem

$$\begin{cases} \operatorname{div}(\mathbb{C}\nabla w) = 0, & \text{in } G, \\ (\mathbb{C}\nabla w)\nu = 0, & \text{on } \Sigma, \\ w = 0, & \text{on } \Sigma. \end{cases} \tag{3.1}$$

From the uniqueness of the solution to the Cauchy problem (see, for instance, Proposition 6.4) and from the weak unique continuation principle (see, for instance, [**45**]), $w \equiv 0$ in G.

Let us prove for instance that $D_2 \subset D_1$. We have that

$$D_2 \setminus \overline{D_1} \subset \Omega \setminus (\overline{D_1 \cup G}), \tag{3.2}$$

$$\partial(\Omega \setminus (\overline{D_1 \cup G})) = \Gamma_1 \cup \Gamma_2, \tag{3.3}$$

where $\Gamma_1 \subset \partial D_1$, $\Gamma_2 = \partial D_2 \cap \partial G$. Now, we need different arguments according to the space dimension.

Step 1: $n = 2$. Let us distinguish the following three cases:

i) $\partial D_1 \cap \Gamma_2$ contains at least two points x_1 and x_2;
ii) $\partial D_1 \cap \Gamma_2 = \{x_1\}$;
iii) $\partial D_1 \cap \Gamma_2 = \emptyset$.

Let us notice that, since u_i satisfies homogeneous Dirichlet conditions on the C^1 boundary ∂D_i, u_i is continuous up to ∂D_i. This result can be obtained, for instance, by adapting arguments by Campanato [**17**] and by Giaquinta and Modica [**27**].

If i) holds, then $u_i(x_j) = 0$, $w(x_j) = 0$ for $i, j = 1, 2$, so that $\overline{r}(x_j) = 0$, for $j = 1, 2$. Recalling (2.8) and (2.9), we have

$$0 = \overline{r}(x_1) - \overline{r}(x_2) = a_3 e_3 \times (x_1 - x_2), \tag{3.4}$$

and, since $x_1 - x_2$ and e_3 are orthogonal nonzero vectors, we have $a_3 = 0$. Hence $\overline{r} \equiv c$, but $\overline{r}(x_1) = 0$ implies $c = 0$. Therefore $\overline{r} \equiv 0$ so that $u_1 \equiv u_2$ in G. By integrating by parts and recalling that $u_1 = 0$ on Γ_1 and $u_1 = u_2 = 0$ on Γ_2, we have

$$\int_{\Omega \setminus (\overline{D_1 \cup G})} \mathbb{C}\nabla u_1 \cdot \nabla u_1 = \int_{\Gamma_1} (\mathbb{C}\nabla u_1)\nu \cdot u_1 + \int_{\Gamma_2} (\mathbb{C}\nabla u_1)\nu \cdot u_1 = 0, \tag{3.5}$$

Hence, recalling also (3.2), we have that $\widehat{\nabla} u_1 \equiv 0$ in $D_2 \setminus \overline{D_1}$. If the open set $D_2 \setminus \overline{D_1}$ were nonempty then, by the weak unique continuation principle, $\widehat{\nabla} u_1 \equiv 0$ in $\Omega \setminus \overline{D_1}$, contradicting the choice of nontrivial φ. Therefore $D_2 \subset \overline{D_1}$ and, since D_2 is open and ∂D_1 is locally graph of a continuous function, $D_2 \subset D_1$.

Let us consider now case ii). It is evident that a path on ∂D_1 connecting a point of Γ_1 with a point of $\partial D_1 \setminus \Gamma_1$ must intersect $\Gamma_2 \cap \partial D_1 = \{x_1\}$. Since $\partial D_1 \setminus \{x_1\}$ is connected and does not intersect $\Gamma_2 \cap \partial D_1$, then it cannot intersect both Γ_1 and $\partial D_1 \setminus \Gamma_1$. Therefore either $\Gamma_1 \subset \{x_1\}$ or $\Gamma_1 \supset \partial D_1 \setminus \{x_1\}$. We may write

$$\int_{\Omega \setminus (\overline{D_1 \cup G})} \mathbb{C}\nabla(u_1 - \overline{r}) \cdot \nabla(u_1 - \overline{r}) = \int_{\Gamma_1} (\mathbb{C}\nabla u_1)\nu \cdot (u_1 - \overline{r}) + \int_{\Gamma_2} (\mathbb{C}\nabla u_1)\nu \cdot (u_1 - \overline{r}). \tag{3.6}$$

The second integral in the right hand side of (3.6) vanishes since $u_1 - \overline{r} = u_2 = 0$ on Γ_2. If $\Gamma_1 \subset \{x_1\}$, then also the first integral vanishes and again, as seen for case i), we have $D_2 \subset D_1$.

If, instead, $\Gamma_1 \supset \partial D_1 \setminus \{x_1\}$, recalling the *equilibrium condition* (1.2), we have

$$\int_{\Gamma_1} (\mathbb{C}\nabla u_1)\nu \cdot (u_1 - \overline{r}) = \int_{\partial D_1} (\mathbb{C}\nabla u_1)\nu \cdot u_1 - \int_{\partial D_1} (\mathbb{C}\nabla u_1)\nu \cdot \overline{r} = 0, \tag{3.7}$$

and again we have $D_2 \subset D_1$.

In case iii), it is easy to see that either $\Gamma_1 = \emptyset$ or $\Gamma_1 = \partial D_1$, and, arguing similarly to case ii), we find again that $\int_{D_2 \setminus \overline{D_1}} \mathbb{C}\nabla u_1 \cdot \nabla u_1 = 0$, and hence $D_2 \subset D_1$.

Step 2: $n = 3$. Let us distinguish the following two cases:

i) $\partial D_1 \cap \Gamma_2$ contains at least three points x_1, x_2 and x_3 not belonging to the same straight line;

ii) $\partial D_1 \cap \Gamma_2$ is contained in a straight line l.

If i) holds, then $u_i(x_j) = 0$, $w(x_j) = 0$ for $i = 1, 2$, $j = 1, 2, 3$, so that $\overline{r}(x_j) = 0$, for $j = 1, 2, 3$. Now,

$$0 = \overline{r}(x_1) - \overline{r}(x_2) = a \times (x_1 - x_2), \qquad 0 = \overline{r}(x_1) - \overline{r}(x_3) = a \times (x_1 - x_3).$$

Since x_1, x_2, x_3 do not belong to the same straight line, we have that $a = 0$. Hence, $\overline{r} \equiv c$, but $\overline{r}(x_1) = 0$ implies $c = 0$. Therefore $\overline{r} \equiv 0$ and $u_1 \equiv u_2$ in G.

By integrating by parts and recalling that $u_1 = 0$ on Γ_1 and $u_1 = u_2 = 0$ on Γ_2, we have

$$\int_{\Omega \setminus (\overline{D_1 \cup G})} \mathbb{C}\nabla u_1 \cdot \nabla u_1 = \int_{\Gamma_1} (\mathbb{C}\nabla u_1)\nu \cdot u_1 + \int_{\Gamma_2} (\mathbb{C}\nabla u_1)\nu \cdot u_1 = 0, \tag{3.8}$$

and, arguing as in the previous step, $D_2 \subset D_1$.

In order to treat case ii), it is useful to introduce the following Lemma, which will be proved at the end of this Chapter.

LEMMA 3.1. *Let D_1 be a domain in $\mathbb{R}^3$ having boundary ∂D_1 connected, of Lipschitz class with constants ρ_0, M_0 and such that* area$(\partial D_1) \le M_2\rho_0^2$. *Given any straight line l, we have that $\partial D_1 \setminus l$ is path connected.*

It is worth noting that, in principle, Definition 2.1 applies to any point P of ∂D_1 with constants ρ_0 and M_0 depending on P. However, by a compactness argument, it is easy to see that, in fact, constants ρ_0 and M_0 satisfying Definition 2.1 can be chosen independently of the point P. Therefore the regularity assumption of the above Lemma is satisfied for suitable parameters ρ_0 and M_0. Moreover, it is straightforward to prove that the estimate $\text{area}(\partial D_1) \leq M_2\rho_0^2$ holds for a suitable constant M_2 only depending on M_0 and $\frac{|D|}{\rho_0^3}$.

It is clear that a path on ∂D_1 connecting a point of Γ_1 with a point of $\partial D_1 \setminus \Gamma_1$ must intersect $\Gamma_2 \cap \partial D_1 \subset l$. Since, by the above lemma, $\partial D_1 \setminus l$ is path connected and it does not intersect $\Gamma_2 \cap \partial D_1$, then it cannot intersect both Γ_1 and $\partial D_1 \setminus \Gamma_1$. Therefore either $\Gamma_1 \subset l$ or $\Gamma_1 \supset \partial D_1 \setminus l$. By arguing similarly to the previous step, we can write again (3.6) and, recalling that a surface integral over a subset of a segment vanishes, we obtain again $D_2 \subset D_1$. □

PROOF OF LEMMA 3.1. Let us notice that $\partial D_1 \setminus l = \partial D_1 \setminus (\partial D_1 \cap l)$ and that $\partial D_1 \cap l$ is a compact subset of the line l, whose connected closure is a closed segment $S \subset l$, with endpoints belonging to $\partial D_1 \cap l$.

Let us fix an orientation on l, which we shall refer to as the positive orientation, so that the segment S will have a starting point P^S and an ending point P^E.

Given any two points P_0, $Q_0 \in \partial D_1 \setminus l$, our aim is to construct a path contained in $\partial D_1 \setminus l$ and joining P_0 and Q_0.

Claim. There exists a constant $K > 0$, only depending on M_0 and M_2, such that any two points P_0 and $Q_0 \in \partial D_1$ can be connected with a path γ contained in ∂D_1 having length

$$\text{length}(\gamma) \leq K\rho_0. \tag{3.9}$$

PROOF OF THE CLAIM. Since, by our assumptions, ∂D_1 is connected and, being locally a continuous graph, also locally path connected, it follows that it is path connected. Therefore, there exists a continuous map $\gamma : [a, b] \to \partial D_1$ joining $P_0 = \gamma(a)$ and $Q_0 = \gamma(b)$.

Let us denote $\beta = \arctan M_0$ and notice that $\cos\beta = (1 + M_0^2)^{-\frac{1}{2}}$. Let us define $\{x_i\}$, $i = 1, ..., s+1$, as follows: $x_1 = P_0$, $x_{s+1} = Q_0$, $x_{i+1} = \gamma(t_i)$, where $t_i = \max\{t \mid |\gamma(t) - x_i| = \rho_0 \cos\beta\}$ if $|x_i - Q_0| > \rho_0\cos\beta$, otherwise let $i = s$ and stop the process. By construction, the balls $B_{\frac{\rho_0 \cos\beta}{2}}(x_i)$ are pairwise disjoint, for $i = 1, ..., s$. Clearly, $\text{area}(\partial D_1 \cap B_{\frac{\rho_0\cos\beta}{2}}(x_i)) \geq \pi\frac{\rho_0^2\cos\beta^2}{4}$, so that $s \leq \frac{4M_2(1+M_0^2)}{\pi}$.

For any $i = 1, ..., s$, by our regularity assumptions, there exists a rigid transformation of coordinates under which we have $x_i = 0$ and

$$\partial D_1 \cap B_{\rho_0}(0) = \{x = (x', x_3) \in B_{\rho_0}(0) \mid x_3 = \psi(x')\}, \tag{3.10}$$

where ψ is a Lipschitz function defined on the disk $B'_{\rho_0}(0)$ in the plane $Ox_1'x_2'$ satisfying

$$\psi(0) = 0, \quad \|\psi\|_{C^{0,1}(B'_{\rho_0}(0))} \leq M_0\rho_0. \tag{3.11}$$

Let us notice that the restriction of the graph of ψ to the disk $\overline{B'_{\rho_0\cos\beta}(0)}$ is contained in ∂D_1.

Let us denote by Π the projection on the plane $Ox_1'x_2'$. Let σ be the rectilinear path having starting point $\Pi(x_i) = 0$ and ending point $\Pi(x_{i+1})$. The path $(\sigma, \psi \circ \sigma)$ joins x_i with x_{i+1} and has length bounded by $\rho_0\cos\beta\sqrt{1 + M_0^2} = \rho_0$. By replacing the path γ in $[t_i, t_{i+1}]$ with $(\sigma, \psi\circ\sigma)$ for any i, $i = 1, ..., s$, we have construct a new path, still denoted by γ, satisfying (3.9) with $K = \frac{4M_2(1+M_0^2)}{\pi}$. □

Now, if $\gamma([a,b]) \subset \partial D_1 \setminus l$, then we are done. Otherwise, let us show how to modify γ to obtain the thesis.

Let us consider the closed, nonempty set

$$J = \{t \in [a,b] \mid \gamma(t) \in \partial D_1 \cap l\}. \tag{3.12}$$

Let us define

$$t_{\min} = \min J, \quad R_{\min} = \gamma(t_{\min}). \tag{3.13}$$

We have that $t_{\min} \in (a,b)$ and $R_{\min} \in \partial D_1 \cap l$.

By considering a local representation of ∂D_1 around $R_{\min}$ of type (3.10), we can choose $\tilde{t}$, $a \le \tilde{t} < t_{\min}$, such that $\gamma([\tilde{t}, t_{\min}]) \subset B_{\rho_0 \cos\beta}(R_{\min})$.

If $|Q_0 - R_{\min}| > \rho_0 \cos\beta$, then let us define

$$t_1 = \max\{t \in (t_{\min}, b] \mid |\gamma(t) - \gamma(t_{\min})| = \rho_0 \cos\beta\},$$

otherwise let us define $t_1 = b$. It is evident that if $t_1 < b$, that is if $\gamma(t_1) \neq Q_0$, the length of $\gamma|_{[t_{\min}, t_1]}$ is at least $\rho_0 \cos\beta$.

Let

$$Z = \{x' \in \overline{B'_{\rho_0 \cos\beta}} \mid (x', \psi(x')) \in \partial D_1 \cap l\}. \tag{3.14}$$

We have that $0 \in Z$ and Z is contained in the closed segment $\Pi(S)$.

If $\overline{B'_{\rho_0 \cos\beta}} \setminus Z$ is connected, then we can obviously construct a path σ joining $(\Pi \circ \gamma)(\tilde{t})$ to $(\Pi \circ \gamma)(t_1)$ inside $\overline{B'_{\rho_0 \cos\beta}} \setminus Z$, except, possibly, for the end point $(\Pi \circ \gamma)(t_1)$. Therefore, the path $(\sigma, \psi \circ \sigma)$ is contained in $\partial D_1 \setminus l$ except, possibly, for $\gamma(t_1)$.

If, otherwise, $\overline{B'_{\rho_0 \cos\beta}} \setminus Z$ is not connected, that is Z is a diameter of $\overline{B'_{\rho_0 \cos\beta}}$, the set $\overline{B'_{\rho_0 \cos\beta}} \setminus Z$ has exactly two connected components.

Let us distinguish two cases: either $(\Pi \circ \gamma)(t_1)$ and $(\Pi \circ \gamma)(\tilde{t})$ belong to the closure of the same connected component, or not.

In the former case we can obviously construct a path σ joining $(\Pi \circ \gamma)(\tilde{t})$ to $(\Pi \circ \gamma)(t_1)$ inside the same connected component, except, possibly, for the endpoint $(\Pi \circ \gamma)(t_1)$. Again, the path $(\sigma, \psi \circ \sigma)$ is contained in $\partial D_1 \setminus l$, except, possibly, for $\gamma(t_1)$.

In the latter case, let us denote by R_1 the end point of the diameter Z with respect to the orientation inherited by the positive orientation introduced on S. We can obviously construct paths σ^+, σ^- inside $\overline{B'_{\rho_0 \cos\beta}} \setminus Z$ joining $(\Pi \circ \gamma)(\tilde{t})$ and $(\Pi \circ \gamma)(t_1)$ with some points P_1 and Q_1, respectively, where the points P_1 and Q_1 satisfy $|P_1 - R_1| = |Q_1 - R_1| = \delta$ and $\delta = \frac{\rho_0 \cos\beta}{2\sqrt{1+M_0^2}}$. The point $(R_1, \psi(R_1))$ belongs to $\partial D_1 \cap l$ and it follows $R_{\min} = \gamma(t_{\min})$ on S with respect to the positive orientation. Moreover, $|(R_1, \psi(R_1)) - R_{\min}| \ge \rho_0 \cos\beta$.

Let us compute

$$\begin{aligned} |(R_1, \psi(R_1)) - (P_1, \psi(P_1))| \le \delta\sqrt{1+M_0^2} = \tfrac{\rho_0 \cos\beta}{2}, \\ |(R_1, \psi(R_1)) - (Q_1, \psi(Q_1))| \le \delta\sqrt{1+M_0^2} = \tfrac{\rho_0 \cos\beta}{2}. \end{aligned} \tag{3.15}$$

Given a local representation of ∂D_1 around $(R_1, \psi(R_1))$ of type (3.10), and still denoting by Π the projection on the plane $Ox'_1x'_2$ relative to this local representation, we have trivially that

$$\begin{aligned} &|\Pi(R_1, \psi(R_1)) - \Pi(P_1, \psi(P_1))| < \rho_0 \cos\beta, \\ &|\Pi(R_1, \psi(R_1)) - \Pi(Q_1, \psi(Q_1))| < \rho_0 \cos\beta. \end{aligned} \tag{3.16}$$

Now, if $\Pi(P_1, \psi(P_1))$ and $\Pi(Q_1, \psi(Q_1))$ belong to the same connected component of $\overline{B'_{\rho_0 \cos\beta}} \setminus Z$, with Z given by (3.14) in the present local representation, then, by following previous arguments, we can join $(P_1, \psi(P_1))$ with $(Q_1, \psi(Q_1))$ inside $\partial D_1 \setminus l$ and therefore, by gluing of paths, we can connect $\gamma(\tilde{t})$ with $\gamma(t_1)$ inside $\partial D_1 \setminus l$ except, possibly, for $\gamma(t_1)$.

Otherwise, if $\Pi(P_1, \psi(P_1))$ and $\Pi(Q_1, \psi(Q_1))$ do not belong to the same connected component of $\overline{B'_{\rho_0 \cos\beta}} \setminus Z$, we can repeat the above construction defining similarly points R_2, P_2, Q_2 and so on. By our regularity assumptions, we have that $\text{length}(S) \leq K\rho_0$, with K only depending on M_0 and M_1. Moreover, at each step, the point $(R_j, \psi(R_j))$ follows $(R_{j-1}, \psi(R_{j-1}))$ on S with respect to the positive orientation of S, with $|(R_j, \psi(R_j)) - (R_{j-1}, \psi(R_{j-1}))| \geq \rho_0 \cos\beta$. Therefore, in a finite number of steps we reduce to the case in which $\Pi(P_k, \psi(P_k))$ and $\Pi(Q_k, \psi(Q_k))$ belong to the same connected component of $\overline{B'_{\rho_0 \cos\beta}} \setminus Z$, for some k, so that, by gluing of paths, we can connect $\gamma(\tilde{t})$ with $\gamma(t_1)$ inside $\partial D_1 \setminus l$ except, possibly, for $\gamma(t_1)$.

Let us still denote by γ the path so modified by this first step.

If $\gamma(t) \in \partial D_1 \setminus l$ for every $t \in [t_1, b]$, then we are done. Otherwise, defining $t_{\min} = \min\{t \in [t_1, b] \mid \gamma(t) \in \partial D_1 \cap l\}$, as a second step we can repeat the above construction starting from the new point $\gamma(t_{\min})$. On the other hand it is evident that for each step of this kind a path of length at least $\rho_0 \cos\beta$ is covered on the given initial path. Therefore, since the length of the given path γ is bounded by (3.9), the numbers of these steps is finite, at most $\left[\frac{K}{\cos\beta}\right] + 1$, so that the points P_0 and Q_0 can be connected inside $\partial D_1 \setminus l$. □

REMARK 3.2. It is reasonable that the statement of Lemma 3.1 continue to hold even removing the Lipschitz regularity assumption. However, we have given here a constructive proof for the case of Lipschitz surfaces since the arguments introduced in this proof represent a preliminary step towards the proof of Lemma 7.1, which is the geometrical result needed in the stability context.

CHAPTER 4

Proof of the stability result

Here and in the sequel we shall denote by G the connected component of the open set $\Omega \setminus \overline{(D_1 \cup D_2)}$ such that $\Sigma \subset \partial G$.

The proof of Theorem 2.5 is obtained from the following sequence of Propositions.

PROPOSITION 4.1 (Lipschitz Propagation of Smallness for the Mixed Problem). *Let Ω be a Lipschitz domain with constants ρ_0, M_0, according to Definition 2.1, and satisfying (2.12). Let D be an open connected subset of Ω satisfying (2.14), (2.16) and of Lipschitz class with constants ρ_0, M_0. Let $u \in H^1(\Omega \setminus \overline{D}, \mathbb{R}^n)$ be the solution to (2.33)-(2.34), where the elasticity tensor $\mathbb{C}$ satisfies (2.24)-(2.26) and the traction field φ satisfies (2.21)- (2.23).*

There exists $s > 1$, only depending on α_0, β_0, M and M_0, such that for every $\rho > 0$ and every $\bar{x} \in (\Omega \setminus \overline{D})_{s\rho}$, we have

$$(4.1) \qquad \int_{B_\rho(\bar{x})} |\widehat{\nabla} u|^2 \geq \frac{C\rho_0}{\exp\left[A\left(\frac{\rho_0}{\rho}\right)^B\right]} \|\varphi\|^2_{H^{-\frac{1}{2}}(\partial\Omega,\mathbb{R}^n)},$$

where $A > 0$, $B > 0$ and $C > 0$ only depend on α_0, β_0, M, M_0, M_1 and F.

PROPOSITION 4.2 (Stability Estimate of Continuation from Cauchy Data). *Let the hypotheses of Theorem 2.5 be satisfied. We have*

$$(4.2) \qquad \int_{D_2\setminus D_1} |\widehat{\nabla} u_1|^2 \leq \rho_0 \|\varphi\|^2_{H^{-\frac{1}{2}}(\partial\Omega,\mathbb{R}^n)} \omega\left(\frac{\epsilon}{\rho_0^{\frac{3-n}{2}} \|\varphi\|_{H^{-\frac{1}{2}}(\partial\Omega,\mathbb{R}^n)}}\right),$$

$$(4.3) \qquad \int_{D_1\setminus D_2} |\widehat{\nabla} u_2|^2 \leq \rho_0 \|\varphi\|^2_{H^{-\frac{1}{2}}(\partial\Omega,\mathbb{R}^n)} \omega\left(\frac{\epsilon}{\rho_0^{\frac{3-n}{2}} \|\varphi\|_{H^{-\frac{1}{2}}(\partial\Omega,\mathbb{R}^n)}}\right),$$

where ω is an increasing continuous function on $[0, \infty)$ which satisfies

$$(4.4) \qquad \omega(t) \leq C(\log|\log t|)^{-c_n}, \qquad \textit{for every } t < e^{-1},$$

with $C > 0$ only depending on α_0, β_0, M, α, M_0, M_1, and $c_n > 0$ only depending on n.

PROPOSITION 4.3 (Improved Stability Estimate of Continuation from Cauchy Data). *Let the hypotheses of Theorem 2.5 hold and, in addition, let us assume that there exist $L > 0$ and $\tilde{\rho}_0$, $0 < \tilde{\rho}_0 \leq \rho_0$, such that ∂G is of Lipschitz class with constants $\tilde{\rho}_0$, L. Then (4.2)-(4.3) hold with ω given by*

$$(4.5) \qquad \omega(t) \leq C|\log t|^{-\gamma}, \qquad \text{for every } t < 1,$$

where $\gamma > 0$ and $C > 0$ only depend on α_0, β_0, M, α, M_0, M_1, L and $\frac{\tilde{\rho}_0}{\rho_0}$.

PROPOSITION 4.4 (Relative Graphs, [4]). *Let Ω be a bounded domain satisfying (2.18) and let D_i, $i = 1, 2$, be two connected open subsets of Ω satisfying (2.14), (2.16) and (2.19). There exist numbers d_0, $\tilde{\rho}_0$, $d_0 > 0$, $0 < \tilde{\rho}_0 \leq \rho_0$, for which the ratios $\frac{d_0}{\rho_0}$, $\frac{\tilde{\rho}_0}{\rho_0}$ only depend on α and M_0, such that if we have*

$$d_{\mathcal{H}}(\overline{D_1}, \overline{D_2}) \leq d_0, \tag{4.6}$$

then every connected component G of $\Omega \setminus (\overline{D_1 \cup D_2})$ has boundary of Lipschitz class with constants $\tilde{\rho}_0$, L, where $\tilde{\rho}_0$ is as above and $L > 0$ only depends on α and M_0.

PROOF OF THEOREM 2.5. Let us denote, for simplicity, $d = d_{\mathcal{H}}(\partial D_1, \partial D_2)$. Let us see that, if $\eta > 0$ is such that

$$\int_{D_2 \setminus D_1} |\widehat{\nabla} u_1|^2 \leq \eta, \qquad \int_{D_1 \setminus D_2} |\widehat{\nabla} u_2|^2 \leq \eta, \tag{4.7}$$

then we have

$$d \leq C\rho_0 \left[\log \left(\frac{C\rho_0 \|\varphi\|^2_{H^{-\frac{1}{2}}(\partial\Omega, \mathbb{R}^n)}}{\eta} \right) \right]^{-\frac{1}{B}}, \tag{4.8}$$

where $B > 0$ and $C > 0$ only depend on α_0, β_0, M, M_0, M_1 and F.

We may assume, with no loss of generality, that there exists $x_0 \in \partial D_1$ such that $\text{dist}(x_0, \partial D_2) = d$. Let us distinguish two cases:

i) $B_d(x_0) \subset D_2$;

ii) $B_d(x_0) \cap D_2 = \emptyset$.

In case i), by the regularity assumptions made on ∂D_1, there exists $x_1 \in D_2 \setminus D_1$ such that $B_{td}(x_1) \subset D_2 \setminus D_1$, with $t = \frac{1}{1+\sqrt{1+M_0^2}}$.

By (4.7) and by Proposition 4.1 with $\rho = \frac{td}{s}$, we have

$$\eta \geq \frac{C\rho_0}{\exp\left[A \left(\frac{s\rho_0}{td} \right)^B \right]} \|\varphi\|^2_{H^{-\frac{1}{2}}(\partial\Omega, \mathbb{R}^n)}, \tag{4.9}$$

where $A > 0$, $B > 0$ and $C > 0$ only depend on α_0, β_0, M, M_0, M_1 and F.

By (4.9) we easily find (4.8).

Case ii) can be treated similarly by substituting u_1 with u_2.

Hence, by Proposition 4.2 and assuming $\epsilon < e^{-e} \rho_0^{\frac{3-n}{2}} \|\varphi\|_{H^{-\frac{1}{2}}(\partial\Omega, \mathbb{R}^n)}$ we obtain

$$d \leq C\rho_0 \left\{ \log \left[\log \left| \log \frac{\epsilon}{\rho_0^{\frac{3-n}{2}} \|\varphi\|_{H^{-\frac{1}{2}}(\partial\Omega, \mathbb{R}^n)}} \right| \right] \right\}^{-\frac{1}{B}}, \tag{4.10}$$

where $B > 0$ and $C > 0$ only depend on α_0, β_0, M, M_0, M_1 and F.

Thus we have obtained a stability estimate of log–log–log type.

Next, by (4.10), we can find $\epsilon_0 > 0$, only depending on α_0, β_0, M, α, M_0, M_1 and F such that if $\epsilon \leq \epsilon_0$ then $d \leq d_0$, where d_0 appears in (4.6). By Proposition 4.4, G satisfies the hypotheses of Proposition 4.3 and therefore the log–log type estimates (2.39), (2.41) follow.

Let us notice that, in general, the Hausdorff distances $d_{\mathcal{H}}(\partial D_1, \partial D_2)$ and $d_{\mathcal{H}}(\overline{D_1}, \overline{D_2})$ are not equivalent. However, in our regularity assumptions, (2.40) can be derived from (2.39) by using estimates contained in the proof of Proposition 3.6 in [**4**].

For a direct proof of (2.40), we can also argue similarly to the proof of (2.39). Let us assume, with no loss of generality, that there exists $x_0 \in \overline{D_1}$ such that $\text{dist}(x_0, \overline{D_2}) = d$. If $B_d(x_0) \subset D_1$, then $B_d(x_0) \subset D_1 \setminus D_2$ and (4.9) follows with t replaced by 1. If, otherwise, $B_d(x_0) \not\subset D_1$, then $\text{dist}(x_0, \partial D_1) \leq d$, and we can distinguish two cases:

i) $\text{dist}(x_0, \partial D_1) > \frac{d}{2}$,

ii) $\text{dist}(x_0, \partial D_1) \leq \frac{d}{2}$.

When i) holds, then $B_{\frac{d}{2}}(x_0) \subset D_1 \setminus D_2$ and again (4.9) follows with t replaced by $\frac{1}{2}$. When ii) holds, there exists $y_0 \in \partial D_1$ such that $|y_0 - x_0| \leq \frac{d}{2}$. Therefore there exists $y_1 \in D_1$ such that $B_{\frac{td}{2}}(y_1) \subset D_1 \setminus D_2$, and (4.9) follows with t replaced by $\frac{t}{2}$. From (4.9), arguing as in the proof of (2.39), we obtain (2.40). □

For the proof of Corollary 2.7, which is based only on geometrical arguments, we refer to [**4**, §4].

CHAPTER 5

Proof of Proposition 4.1

The proof of Proposition 4.1 is essentially based on the following Proposition, which was obtained in [**36**, Proposition 3.1].

PROPOSITION 5.1 (Lipschitz Propagation of Smallness for the Neumann Problem). *Let U be a bounded domain in $\mathbb{R}^n$ with boundary of Lipschitz class with constants ρ_0, M_0, according to Definition 2.1, and satisfying* (2.12). *Let $u \in H^1(U, \mathbb{R}^n)$ be any solution to*

$$\begin{cases} \operatorname{div}(\mathbb{C}\nabla u) = 0, & \text{in } U, \\ (\mathbb{C}\nabla u)\nu = \tilde{\varphi}, & \text{on } \partial U, \end{cases} \tag{5.1}$$

where $\mathbb{C}$ satisfies (2.24)–(2.26) *and $\tilde{\varphi}$ satisfies*

$$\tilde{\varphi} \in H^{-\frac{1}{2}}(\partial U, \mathbb{R}^n), \quad \tilde{\varphi} \not\equiv 0, \tag{5.2}$$

$$\int_{\partial U} \tilde{\varphi} \cdot r = 0, \quad \text{for every } r \in \mathcal{R}, \tag{5.3}$$

$$\frac{\|\tilde{\varphi}\|_{H^{-\frac{1}{2}}(\partial U, \mathbb{R}^n)}}{\|\tilde{\varphi}\|_{H^{-1}(\partial U, \mathbb{R}^n)}} \leq \tilde{F}, \tag{5.4}$$

where $\tilde{F} > 0$ is a given constant. There exists $s > 1$, only depending on α_0, β_0, M and M_0, such that for every $\rho > 0$ and every $\bar{x} \in U_{s\rho}$, we have

$$\int_{B_\rho(\bar{x})} |\widehat{\nabla} u|^2 \geq \frac{C}{\exp\left[A\left(\frac{\rho_0}{\rho}\right)^B\right]} \int_U |\widehat{\nabla} u|^2, \tag{5.5}$$

where $A > 0$, $B > 0$ and $C > 0$ only depend on α_0, β_0, M, M_0, M_1 and $\tilde{F}$.

REMARK 5.2. The solution of problem (5.1) is determined up to an infinitesimal rigid displacement.

For the proof of Proposition 4.1 we need also the following auxiliary propositions.

PROPOSITION 5.3 (Korn–type Inequality). *Let U be a bounded domain in $\mathbb{R}^n$ with boundary of Lipschitz class with constants ρ_0, M_0 and satisfying* (2.12). *For every $u \in H^1(U, \mathbb{R}^n)$ such that*

$$u = 0, \quad \text{on } \partial U \cap B_{\rho_0}(P_1), \tag{5.6}$$

where P_1 is some point in ∂U, we have

$$\int_U |\nabla u|^2 \leq C \int_U |\widehat{\nabla} u|^2, \tag{5.7}$$

where C is a positive constant only depending on M_0 and M_1.

PROPOSITION 5.4 (Poincaré–type Inequality). *Let U be a bounded domain in $\mathbb{R}^n$ with boundary of Lipschitz class with constants ρ_0, M_0 and satisfying (2.12). For every $u \in H^1(U, \mathbb{R}^n)$ such that*

$$u = 0, \quad on\ \partial U \cap B_{\rho_0}(P_1), \tag{5.8}$$

where P_1 is some point in ∂U, we have

$$\int_U |u|^2 \le C \int_U |\nabla u|^2, \tag{5.9}$$

where C is a positive constant only depending on M_0 and M_1.

In order to prove Proposition 5.3 we shall use two constructive Korn–type inequalities due to Kondrat'ev and Oleinik [**32**] (Proposition 5.5 and Proposition 5.6 below).

PROPOSITION 5.5. ([**32**], Theorem 1) *Let U be a bounded domain in $\mathbb{R}^n$ which is starlike with respect to the ball B_{R_1}. For every $u \in H^1(U, \mathbb{R}^n)$ we have*
(5.10)

$$\int_U |\nabla u|^2 \le C_2 \left(\frac{\mathrm{diam(U)}}{R_1}\right)^2 \left(\log\left(\frac{\mathrm{diam(U)}}{R_1}\right) \int_U |\widehat{\nabla} u|^2 + \int_{B_{R_1}} |\nabla u|^2\right), n = 2,$$

$$\int_U |\nabla u|^2 \le C_3 \left(\frac{\mathrm{diam(U)}}{R_1}\right)^3 \left(\int_U |\widehat{\nabla} u|^2 + \int_{B_{R_1}} |\nabla u|^2\right), \quad n = 3, \tag{5.11}$$

where C_2 and C_3 are positive constants only depending on the dimension $n = 2$ or $n = 3$, respectively.

PROPOSITION 5.6. ([**32**], Theorem 2) *Let*

$$C_{l',l} = \{x = (x', x_n) \in \mathbb{R}^n \mid |x'| < l', \ -l < x_n < l\}, \tag{5.12}$$

where $l > l'$. For every $u \in H^1(C_{l',l}, \mathbb{R}^n)$ such that $u = 0$ on $\{x_n = -l\}$, we have

$$\int_{C_{l',l}} |\nabla u|^2 \le C \left(1 + \frac{4l^2}{l'^2}\right) \int_{C_{l',l}} |\widehat{\nabla} u|^2, \tag{5.13}$$

where $C > 0$ is a positive constant only depending on the dimension n.

PROOF OF PROPOSITION 5.3. Let us tessellate $\mathbb{R}^n$ with internally nonoverlapping closed cubes of side $2r$, with r to be chosen later on. Let Q_1,..., Q_N be those cubes which intersect $\overline{U}$, where the cube Q_1 contains the point P_1 appearing in (5.6). For any j, $j = 1, ..., N$, let us denote by $\tilde{Q}_j$ the cube obtained dilating Q_j by a factor 2. If $\tilde{Q}_j \subset U$ then let us define $U_j = \mathrm{int}(\tilde{Q}_j)$. Notice that if Q_i, Q_j have at least a common point and their dilated $\tilde{Q}_i$, $\tilde{Q}_j$ are contained in U, then $U_i \cap U_j = \tilde{Q}_i \cap \tilde{Q}_j$ contains a ball of radius r. Therefore both U_i and U_j are starlike with respect to the same ball of radius r. Let us notice also that if we assume that $4\sqrt{n}r \le \frac{\rho_0}{3\sqrt{1+M_0^2}}$ then any cube Q_j having nonempty intersection with $U_{\frac{\rho_0}{3\sqrt{1+M_0^2}}}$ is such that $\tilde{Q}_j \subset U$.

Now, let Q_i be a cube such that $\tilde{Q}_i$ is not contained in U. Then $\tilde{Q}_i \cap \partial U \neq \emptyset$. Let us choose $P_i \in \tilde{Q}_i \cap \partial U$, with P_1 the same point appearing in (5.6). Given the local representation of U near P_i, as stated in Definition 2.1,

$$U \cap B_{\rho_0}(P_i) = \{x = (x', x_n) \in B_{\rho_0}(0) \quad | \quad x_n > \psi(x')\},$$

let us define

$$U_i = \{x = (x', x_n) \mid |x'| < \rho, \psi(x') < x_n < \frac{\sqrt{3}}{2}\rho_0\}, \tag{5.14}$$

with ρ a positive constant to be chosen later on and satisfying $\rho \leq \frac{\rho_0}{2}$, so that $U_i \subset U$. Our aim is to choose ρ small enough to ensure that U_i is starlike with respect to some ball contained in U_i and then to choose r small enough to ensure that $\tilde{Q}_i \cap U \subset U_i$. If $6\rho M_0 \leq \frac{\sqrt{3}}{2}\rho_0$, then U_i is starlike with respect to the cylinder $\{|x'| \leq \rho, 3\rho M_0 < x_n < \frac{\sqrt{3}}{2}\rho_0\}$, having radius ρ and height $h = \frac{\sqrt{3}}{2}\rho_0 - 3\rho M_0 \geq 3\rho M_0$ and if $4r\sqrt{n} \leq \rho$, then $\tilde{Q}_i \cap U \subset U_i$. Therefore let $\rho = \rho_0 \min\{\frac{1}{2}, \frac{\sqrt{3}}{12 M_0}\}$ and $r = \min\{\frac{\rho_0}{12\sqrt{n(1+M_0^2)}}, \frac{\rho}{4\sqrt{n}}\} = \rho_0 \min\{\frac{1}{12\sqrt{n(1+M_0^2)}}, \frac{\sqrt{3}}{48\sqrt{n}M_0}\}$.

Let $P_i^* = (x' = 0, x_n = \frac{\rho_0}{2})$. It is easy to see that U_i is starlike with respect to the ball $B_s(P_i^*)$, with $s = \min\{\rho, \frac{\sqrt{3}-1}{2}\rho_0, \frac{\rho_0}{2} - 3\rho M_0\} \geq \min\{\rho, \frac{2-\sqrt{3}}{4}\rho_0\}$, where we have taken into account the definition of ρ. Moreover, we have that $\operatorname{dist}(P_i^*, \partial U) \geq \frac{\rho_0}{2\sqrt{1+M_0^2}}$ so that $P_i^* \in U_{\frac{\rho_0}{3\sqrt{1+M_0^2}}}$. Let $Q_{j(i)}$ be a cube in the collection containing P_i^*. Recalling that $U_{j(i)} = \tilde{Q}_{j(i)}$, we have that $U_{j(i)}$ is starlike with respect to the ball $B_r(P_i^*)$. By the choices made for ρ and r, it follows that both U_i and $U_{j(i)}$ are starlike with respect to the ball $B_{R_1}(P_i^*)$, where $R_1 = \rho_0 \min\{\frac{2-\sqrt{3}}{4}, \frac{\sqrt{3}}{48\sqrt{n}M_0}, \frac{1}{12\sqrt{n(1+M_0^2)}}\}$.

We obviously have that $U = \cup_{j=1}^N U_j$. Our aim is to prove that, given any U_j, $j = 2, ..., N$, there exists a finite sequence U_{k_h}, $h = 1, ..., m$, such that $k_h \neq k_{h'}$, for $h \neq h'$, $U_{k_1} = U_j$, $U_{k_m} = U_1$, U_{k_h} is starlike with respect to a ball of radius R_1 which is contained in $U_{k_{h+1}}$, $h = 1, ..., m-1$. Let $J = \{j \mid Q_j \cap U_{\frac{\rho_0}{3\sqrt{1+M_0^2}}} \neq \emptyset\}$. Since $U_{\frac{\rho_0}{3\sqrt{1+M_0^2}}}$ is connected, then also $\cup_{j\in J} Q_j$ is connected. Therefore, if $j \in J$, then we can trivially construct a finite sequence of different cubes Q_{k_h}, $h = 1, ..., m$, $k_h \in J$, such that $Q_{k_1} = Q_j$, $Q_{k_m} = Q_{j(1)}$ and Q_{k_h} and $Q_{k_{h+1}}$ have at least a common point. Then, letting $k_{m+1} = 1$, the sequence U_{k_h}, $h = 1, ..., m+1$, satisfies the above requirements.

If $j \notin J$, then, since, being U connected, also $\cup_{j=1}^N Q_j$ is connected, we can similarly construct a finite sequence of different cubes Q_{k_h}, with $h = 1, ..., m$ and $k_h \in \{1, ..., N\}$, such that $Q_{k_1} = Q_j$, $Q_{k_m} = Q_1$ and Q_{k_h} and $Q_{k_{h+1}}$ have at least a common point. Now, if for every $h = 1, ..., m-1$, $\tilde{Q}_{k_h} \subset U$ then, as above, we are done. Otherwise, let k_{h^*} be the first index such that $\tilde{Q}_{k_{h^*}} \not\subset U$. We know that $U_{k_{h^*}}$ is starlike with respect to a ball of radius R_1 contained in $U_{j(k_{h^*})} = \tilde{Q}_{j(k_{h^*})}$. We can continue the sequence by inserting, as above seen, a sequence U_{k_l}, $l = h^* + 1, ..., L$, $k_l \in J$, for $l = h^* + 1, ..., L-1$, $k_{L-1} = j(1)$, $k_L = 1$, obtaining the desired result.

In the local representation of U_1 near P_1, we have that

$$U_1 \subset C_{\rho, \frac{\sqrt{3}}{2}\rho_0} = \{|x'| < \rho, -\frac{\sqrt{3}}{2}\rho_0 < x_n < \frac{\sqrt{3}}{2}\rho_0\}.$$

Let us define

$$u^0 = \begin{cases} u & \text{in } U_1, \\ 0 & \text{in } C_{\rho,\frac{\sqrt{3}}{2}\rho_0} \setminus U_1. \end{cases} \tag{5.15}$$

By (5.6), the map u_0 belongs to $H^1(C_{\rho,\frac{\sqrt{3}}{2}\rho_0}, \mathbb{R}^n)$ and, therefore, it satisfies the hypotheses of Proposition 5.6. Hence we have

$$\int_{U_1} |\nabla u|^2 \leq C \int_{U_1} |\widehat{\nabla} u|^2, \tag{5.16}$$

where C only depends on M_0. If $j > 1$, let us consider a finite sequence of domains U_{k_h}, $h = 1, ..., m$, as constructed above.

By Proposition 5.5, for every $h = 1, ..., m-1$

$$\int_{U_{k_h}} |\nabla u|^2 \leq C \left(\int_{U_{k_h}} |\widehat{\nabla} u|^2 + \int_{U_{k_{h+1}}} |\nabla u|^2 \right) \leq \leq C \left(\int_{U} |\widehat{\nabla} u|^2 + \int_{U_{k_{h+1}}} |\nabla u|^2 \right), \tag{5.17}$$

with C only depending on M_0. By iterating this inequality over $h = 1, ..., m-1$, and by (5.16), we have

$$\int_{U_j} |\nabla u|^2 \leq C \int_{U} |\widehat{\nabla} u|^2, \tag{5.18}$$

with C only depending on M_0 and m. Since $m \leq N$ and $U = \cup_{j=1}^{N} U_j$, we obtain

$$\int_{U} |\nabla u|^2 \leq C \int_{U} |\widehat{\nabla} u|^2, \tag{5.19}$$

with C only depending on M_0 and N. On the other hand, one can show (see for instance [**9**, Lemma 2.8]) that

$$N \leq C \frac{|U|}{r^n}, \tag{5.20}$$

where $C > 0$ only depends on M_0. By (2.12), and by the choice of r, we can dominate N with a constant only depending on M_1, so that the thesis follows. □

Proof of Proposition 5.4. A proof of this proposition can be obtained by adapting the arguments used in the proof of [**7**, Proposition 3.2]. □

Proof of Proposition 4.1. By applying the Proposition 5.1 to the solution $u \in H^1(\Omega \setminus \overline{D}, \mathbb{R}^n)$ of problem (2.33), (2.34), we have that there exists $s > 1$, only depending on α_0, β_0, M, M_0, such that for every $\rho > 0$ and every $\bar{x} \in (\Omega \setminus \overline{D})_{s\rho}$,

$$\int_{B_\rho(\bar{x})} |\widehat{\nabla} u|^2 \geq \frac{C}{\exp\left[A\left(\frac{\rho_0}{\rho}\right)^B\right]} \int_{\Omega \setminus \overline{D}} |\widehat{\nabla} u|^2, \tag{5.21}$$

where $A>0$, $B>0$ and $C>0$ only depend on α_0, β_0, M, M_0, M_1 and $\tilde{F}$. Here $\tilde{F}$ is an upper bound for $\frac{\|\tilde{\varphi}\|_{H^{-\frac{1}{2}}(\partial(\Omega\setminus\overline{D}),\mathbb{R}^n)}}{\|\tilde{\varphi}\|_{H^{-1}(\partial(\Omega\setminus\overline{D}),\mathbb{R}^n)}}$ and $\tilde{\varphi}$ is defined by

$$\tilde{\varphi}=\begin{cases}\varphi & \text{on } \partial\Omega,\\ (\mathbb{C}\nabla u)\nu & \text{on } \partial D.\end{cases} \tag{5.22}$$

Let $u^0\in H^1(\Omega,\mathbb{R}^n)$ be defined by

$$u^0=\begin{cases}u & \text{in } \Omega\setminus\overline{D},\\ 0 & \text{in } \overline{D}.\end{cases} \tag{5.23}$$

By applying Proposition 5.3, Proposition 5.4 and by standard trace inequalities we have

$$\int_{\Omega\setminus\overline{D}}|\widehat{\nabla} u|^2\geq\frac{C}{\rho_0^2}\|u\|^2_{H^1(\Omega\setminus\overline{D},\mathbb{R}^n)}=\frac{C}{\rho_0^2}\|u^0\|^2_{H^1(\Omega,\mathbb{R}^n)}\geq C\rho_0\|\varphi\|^2_{H^{-\frac{1}{2}}(\partial\Omega,\mathbb{R}^n)}, \tag{5.24}$$

with C only depending on M_0 and M_1. Due to the normalization condition (1.3), we can prove that

$$\|u\|_{H^1(\Omega\setminus\overline{D},\mathbb{R}^n)}\leq C\rho_0^{\frac{3}{2}}\|\tilde{\varphi}\|_{H^{-\frac{1}{2}}(\partial(\Omega\setminus\overline{D}),\mathbb{R}^n)}, \tag{5.25}$$

with C only depending on M_0 and M_1 and ξ_0. In fact, by standard trace inequalities, by Proposition 5.3, Proposition 5.4 and by the strong convexity assumptions (2.30)

$$\begin{aligned}\|u\|^2_{H^1(\Omega\setminus\overline{D},\mathbb{R}^n)}&\leq C\rho_0^2\int_{\Omega\setminus\overline{D}}|\widehat{\nabla} u|^2\leq C\rho_0^2\int_{\Omega\setminus\overline{D}}\mathbb{C}\nabla u\cdot\nabla u=C\rho_0^2\int_{\partial(\Omega\setminus\overline{D})}\tilde{\varphi}\cdot u\leq\\ &\leq C\rho_0^2\|u\|_{H^{\frac{1}{2}}(\partial(\Omega\setminus\overline{D}),\mathbb{R}^n)}\|\tilde{\varphi}\|_{H^{-\frac{1}{2}}(\partial(\Omega\setminus\overline{D}),\mathbb{R}^n)}\leq\\ &\leq C\rho_0^{\frac{3}{2}}\|u\|_{H^1(\Omega\setminus\overline{D},\mathbb{R}^n)}\|\tilde{\varphi}\|_{H^{-\frac{1}{2}}(\partial(\Omega\setminus\overline{D}),\mathbb{R}^n)},\end{aligned}$$

with the stated dependence for C.

Again by standard trace inequalities, Proposition 5.3, Proposition 5.4, by (2.30) and by (5.25), we obtain

$$\begin{aligned}\|\tilde{\varphi}\|^2_{H^{-\frac{1}{2}}(\partial(\Omega\setminus\overline{D}),\mathbb{R}^n)}&\leq\frac{C}{\rho_0^3}\|u\|^2_{H^1(\Omega\setminus\overline{D},\mathbb{R}^n)}\leq\frac{C}{\rho_0}\int_{\Omega\setminus\overline{D}}\mathbb{C}\nabla u\cdot\nabla u=\frac{C}{\rho_0}\int_{\partial\Omega}\varphi\cdot u\leq\\ &\leq\frac{C}{\rho_0}\|\varphi\|_{H^{-\frac{1}{2}}(\partial\Omega,\mathbb{R}^n)}\|u\|_{H^{\frac{1}{2}}(\partial\Omega,\mathbb{R}^n)}\leq\frac{C}{\rho_0^{\frac{3}{2}}}\|\varphi\|_{H^{-\frac{1}{2}}(\partial\Omega,\mathbb{R}^n)}\|u^0\|_{H^1(\Omega,\mathbb{R}^n)}=\\ &=\frac{C}{\rho_0^{\frac{3}{2}}}\|\varphi\|_{H^{-\frac{1}{2}}(\partial\Omega,\mathbb{R}^n)}\|u\|_{H^1(\Omega\setminus\overline{D},\mathbb{R}^n)}\leq\|\varphi\|_{H^{-\frac{1}{2}}(\partial\Omega,\mathbb{R}^n)}\|\tilde{\varphi}\|_{H^{-\frac{1}{2}}(\partial(\Omega\setminus\overline{D}),\mathbb{R}^n)},\end{aligned}$$

so that

$$\|\tilde{\varphi}\|_{H^{-\frac{1}{2}}(\partial(\Omega\setminus\overline{D}),\mathbb{R}^n)}\leq\|\varphi\|_{H^{-\frac{1}{2}}(\partial\Omega,\mathbb{R}^n)}, \tag{5.26}$$

with C only depending on M_0 and M_1 and ξ_0.

Given any $w\in H^1(\partial\Omega,\mathbb{R}^n)$, let us consider $w^0\in H^1(\partial(\Omega\setminus\overline{D}),\mathbb{R}^n)$ defined as follows

$$w^0=\begin{cases}w & \text{on } \partial\Omega,\\ 0 & \text{on } \partial D.\end{cases}$$

Notice that $\|w^0\|_{H^1(\partial(\Omega\setminus\overline{D}),\mathbb{R}^n)} = \|w\|_{H^1(\partial\Omega,\mathbb{R}^n)}$. We can compute

(5.27)

$$\|\varphi\|_{H^{-1}(\partial\Omega,\mathbb{R}^n)} = \sup_{\substack{w\in H^1(\partial\Omega,\mathbb{R}^n)\\ w\neq 0}} \frac{\int_{\partial\Omega}\varphi\cdot w}{\|w\|_{H^1(\partial\Omega,\mathbb{R}^n)}} = \sup_{\substack{w\in H^1(\partial\Omega,\mathbb{R}^n)\\ w\neq 0}} \frac{\int_{\partial(\Omega\setminus\overline{D})}\tilde{\varphi}\cdot w^0}{\|w^0\|_{H^1(\partial(\Omega\setminus\overline{D}),\mathbb{R}^n)}} \leq$$

$$\leq \sup_{\substack{v\in H^1(\partial(\Omega\setminus\overline{D}),\mathbb{R}^n)\\ v\neq 0}} \frac{\int_{\partial(\Omega\setminus\overline{D})}\tilde{\varphi}\cdot v}{\|v\|_{H^1(\partial(\Omega\setminus\overline{D}),\mathbb{R}^n)}} = \|\tilde{\varphi}\|_{H^{-1}(\partial(\Omega\setminus\overline{D}),\mathbb{R}^n)}.$$

From (5.26) and (5.27) it follows that

$$\tilde{F} \leq CF, \tag{5.28}$$

with C only depending on M_0 and M_1. Finally, from (5.21), (5.24) and (5.28) the thesis follows. □

CHAPTER 6

Stability estimates of continuation from Cauchy data

Throughout this Chapter, let Ω be a domain satisfying (2.12) and (2.18). Let D_i, $i=1,2$, be two connected open subsets of Ω satisfying (2.14), (2.16) and (2.19) for $D=D_i$, $i=1,2$. In order to simplify the notation, let us denote

$$\Omega_i = \Omega \setminus \overline{D_i}, \quad \text{i=1,2.} \tag{6.1}$$

Notice that $\overline{\Omega_1 \cap \Omega_2} = \overline{\Omega} \setminus (D_1 \cup D_2)$.

We denote

$$\mathcal{U}^\rho = \{x \in \overline{\Omega} \text{ s.t. } \operatorname{dist}(x, \partial\Omega) \leq \rho\}, \quad \text{for } \rho < \rho_0.$$

The regularity estimates stated in the Lemma below hold for a general strongly convex elasticity tensor $\mathbb{C}$ (not necessarily of Lamé type), satisfying some mild regularity assumptions.

LEMMA 6.1. *Let the domain Ω_i, $i=1,2$, be as above. Let the elasticity tensor $\mathbb{C} \in C^{0,\alpha}(\mathbb{R}^n, \mathcal{L}(\mathbb{M}^n, \mathbb{M}^n))$ satisfy (2.27)–(2.30). Let $u_i \in H^1(\Omega_i, \mathbb{R}^n)$ be the weak solution to the mixed problem (2.33), (2.34), when $D = D_i$, $i=1,2$, where φ satisfies (2.21) and (2.22). Then $u_i \in C^{1,\alpha}(\overline{\Omega_i \setminus \mathcal{U}^{\frac{\rho_0}{8}}}, \mathbb{R}^n)$ and*

$$\|u_i\|_{C^{1,\alpha}(\overline{\Omega_i \setminus \mathcal{U}^{\frac{\rho_0}{8}}}, \mathbb{R}^n)} \leq C\rho_0^{\frac{3-n}{2}} \|\varphi\|_{H^{-\frac{1}{2}}(\partial\Omega, \mathbb{R}^n)}, \quad \text{for } i=1,2, \tag{6.2}$$

$$\|u_1 - u_2\|_{C^{1,\alpha}(\overline{\Omega_1 \cap \Omega_2}, \mathbb{R}^n)} \leq C\rho_0^{\frac{3-n}{2}} \|\varphi\|_{H^{-\frac{1}{2}}(\partial\Omega, \mathbb{R}^n)} \tag{6.3}$$

where $C > 0$ only depends on ξ_0, α, M_0, M_1 and $\|\mathbb{C}\|_{C^{0,\alpha}(\mathbb{R}^n, \mathcal{L}(\mathbb{M}^n, \mathbb{M}^n))}$.

PROOF. Since $u_i = 0$ on ∂D_i, $i=1,2$, by adapting arguments about regularity estimates up to the boundary for solutions to elliptic systems satisfying homogeneous Dirichlet conditions (see, for instance, [**17**], [**18**] and [**27**]), we obtain

$$\|u_i\|_{C^{1,\alpha}(\overline{\Omega_i \setminus \mathcal{U}^{\frac{\rho_0}{8}}}, \mathbb{R}^n)} \leq \frac{C}{\rho_0^{\frac{n}{2}}} \|u_i\|_{H^1(\Omega_i, \mathbb{R}^n)}, \tag{6.4}$$

where $C > 0$ only depends on ξ_0, α, M_0, M_1 and $\|\mathbb{C}\|_{C^{0,\alpha}(\mathbb{R}^n, \mathcal{L}(\mathbb{M}^n, \mathbb{M}^n))}$. Moreover, by applying the inequalities (5.25), (5.26), the following global estimate holds

$$\|u_i\|_{H^1(\Omega_i, \mathbb{R}^n)} \leq C\rho_0^{\frac{3}{2}} \|\varphi\|_{H^{-\frac{1}{2}}(\partial\Omega, \mathbb{R}^n)}, \tag{6.5}$$

where $C > 0$ only depends on ξ_0, M_0 and M_1.

From (6.4) and (6.5), (6.2) follows.

Similarly, by adapting arguments about regularity estimates up to the boundary for solutions to elliptic systems satisfying homogeneous Neumann conditions

$(\mathbb{C}\nabla(u_1-u_2))\nu = 0$ on $\partial\Omega$, (see, for instance, **[17]**, **[18]** and **[27]**), the $C^{1,\alpha}$ norm of $u_1 - u_2$ can be estimated in $\mathcal{U}^{\frac{\rho_0}{2}}$, and then, by using also (6.2), we obtain (6.3). □

LEMMA 6.2. *Under the hypotheses of Theorem 2.5, we have*

$$\|\overline{r}\|_{L^\infty(\overline{\Omega},\mathbb{R}^n)} \leq C\rho_0^{\frac{3-n}{2}} \|\varphi\|_{H^{-\frac{1}{2}}(\partial\Omega,\mathbb{R}^n)}, \tag{6.6}$$

where $\overline{r}$ *is the infinitesimal rigid displacement appearing in* (2.38) *and* $C > 0$ *only depends on* α_0, β_0, M, α, M_0 *and* M_1.

PROOF. Let $v = u_1 - u_2$, with $u_i \in H^1(\Omega_i, \mathbb{R}^n)$ the weak solution to the mixed problem (2.33), (2.34), when $D = D_i$, $i = 1, 2$, where φ satisfies (2.21) and (2.22).

The infinitesimal rigid displacement $\overline{r}$ appearing in (2.38) and minimizing the distance in $L^2(\Sigma)$ of $u_1 - u_2$ from $\mathcal{R}$, is given by

$$\overline{r}(x) = c + a \times x, \quad x \in \overline{\Omega}, \tag{6.7}$$

with the n-vectors c, a which solve the linear system

$$\begin{cases} c|\Sigma| + a \times S_\Sigma(0) = \int_\Sigma v, \\ c \times S_\Sigma(0) - I_\Sigma(0)a = -\int_\Sigma x \times v \end{cases} \tag{6.8}$$

and where 0 is the origin of a coordinate system in $\mathbb{R}^n$, to be chosen later on. In (6.8),

$$S_\Sigma(0) = \int_\Sigma x \tag{6.9}$$

and

$$I_\Sigma(0) = \int_\Sigma \left(|x|^2 I_n - x \otimes x\right) \tag{6.10}$$

are, respectively, the n-vector of the *first order moments* of Σ and the $n \times n$ *inertia tensor* of Σ evaluated with respect to 0. By choosing the origin 0 coincident with the centre of mass G_Σ of Σ, we have

$$S_\Sigma(G_\Sigma) = 0, \quad I_\Sigma(G_\Sigma) = \int_\Sigma \left(|x - G_\Sigma|^2 I_n - (x - G_\Sigma) \otimes (x - G_\Sigma)\right) \tag{6.11}$$

and

$$\overline{r}(x) = c^* + a \times (x - G_\Sigma), \quad x \in \overline{\Omega}, \tag{6.12}$$

with $c^* = c + a \times G_\Sigma$. The vector c^* is given by

$$c^* = \frac{1}{|\Sigma|}\int_\Sigma v. \tag{6.13}$$

Moreover, since $I_\Sigma(G_\Sigma)$ is a positive definite tensor, the vector a is the unique solution of the equation

$$I_\Sigma(G_\Sigma)a = \int_\Sigma (x - G_\Sigma) \times v. \tag{6.14}$$

Note that G_Σ is an internal point of the smallest closed convex set containing Σ and

$$|G_\Sigma - x| \leq C\rho_0 \quad \text{for every } x \in \overline{\Omega}, \tag{6.15}$$

where $C > 0$ only depends on α, M_0 and M_1.

At this point, it is convenient to treat separately the cases $n = 2$ and $n = 3$.

Case n=2. By (2.9), (6.14) becomes

$$a_3 = \left(I^0_\Sigma(G_\Sigma)\right)^{-1} \int_\Sigma (x - G_\Sigma) \times v \cdot e_3, \tag{6.16}$$

where

$$I^0_\Sigma(G_\Sigma) = \int_\Sigma |x - G_\Sigma|^2 \tag{6.17}$$

is the *polar moment of inertia* of Σ evaluated with respect to G_Σ.

In order to estimate a_3, we shall prove that there exists a constant $C > 0$ such that

$$I^0_\Sigma(G_\Sigma) \geq C\rho_0^3, \tag{6.18}$$

with C only depending on M_0. By our assumption (2.17) on the point $P_0 \in \Sigma$ and by the regularity assumption (2.20) on Σ we have

$$I^0_\Sigma(G_\Sigma) \geq \int_{B_{\rho_0 \cos\beta}(P_0)\cap\Sigma} |x - G_\Sigma|^2, \tag{6.19}$$

where $\tan\beta = M_0$. By the local representation of Σ near P_0 and by expressing the integrand in (6.19) in terms the coordinate system (x', x_n) with origin at P_0, we have

$$\begin{aligned}\int_{B_{\rho_0 \cos\beta}(P_0)\cap\Sigma} |x - G_\Sigma|^2 &= \\ = \int_{-\rho_0\cos\beta}^{\rho_0\cos\beta} &\left((x' - x'(G_\Sigma))^2 + (\psi(x') - x_n(G_\Sigma))^2\right)\sqrt{1 + (\psi'(x'))^2}dx' \geq \\ \geq \int_{-\rho_0\cos\beta}^{\rho_0\cos\beta} &(x' - x'(G_\Sigma))^2 dx' \geq \min_{t\in\mathbb{R}} \int_{-\rho_0\cos\beta}^{\rho_0\cos\beta} (x' - t)^2 dx' = \\ &= \int_{-\rho_0\cos\beta}^{\rho_0\cos\beta} x'^2 dx' = \frac{2}{3}\rho_0^3 \cos^3\beta.\end{aligned} \tag{6.20}$$

By inserting (6.20) in (6.19), the inequality (6.18) follows.

By (6.16), Hölder inequality and (6.18) we have

$$|a_3| \leq \left(I^0_\Sigma(G_\Sigma)\right)^{-\frac{1}{2}} \|v\|_{L^2(\Sigma,\mathbb{R}^n)} \leq \frac{C}{\rho_0^{\frac{3}{2}}} \|v\|_{L^2(\Sigma,\mathbb{R}^n)}, \tag{6.21}$$

where $C > 0$ only depends on M_0.

Again by our regularity assumptions on the boundary we have

$$|\Sigma| \geq |B_{\rho_0}(P_0) \cap \partial\Omega| \geq 2\rho_0 \cos\beta, \tag{6.22}$$

and, therefore, by (6.13) and by Hölder inequality we have

$$|c^*| \leq \frac{1}{|\Sigma|^{\frac{1}{2}}} \|v\|_{L^2(\Sigma,\mathbb{R}^n)} \leq \frac{C}{\rho_0^{\frac{1}{2}}} \|v\|_{L^2(\Sigma,\mathbb{R}^n)}, \tag{6.23}$$

where $C > 0$ only depends on M_0.

Finally, by the estimates (6.21) and (6.23) for a_3 and c^*, respectively, by (6.15), by a trace inequality and by the global estimate (6.5), we have, for any $x \in \overline{\Omega}$,

$$(6.24)\quad |\overline{r}(x)| = |c^* + a \times (x - G_\Sigma)| \leq |c^*| + |a_3| \max_{x\in\overline{\Omega}} |x - G_\Sigma| \leq$$
$$\leq \frac{C}{\rho_0^{\frac{1}{2}}}\|v\|_{L^2(\Sigma,\mathbb{R}^n)} + \frac{C}{\rho_0^{\frac{3}{2}}}\|v\|_{L^2(\Sigma,\mathbb{R}^n)}\rho_0 \leq C\rho_0\|\varphi\|_{H^{-\frac{1}{2}}(\partial\Omega,\mathbb{R}^n)},$$

where $C > 0$ only depends on α_0, β_0, M, α, M_0 and M_1.

Case n=3. Let $\{v^{(i)}\}_{i=1}^3$ be the orthonormal set of eigenvectors of $I_\Sigma(G_\Sigma)$ and let $\{\lambda_i\}_{i=1}^3$ be the corresponding eigenvalues, with $0 < \lambda_1 \leq \lambda_2 \leq \lambda_3$. Then, we can write

$$(6.25)\qquad I_\Sigma(G_\Sigma) = \sum_{i=1}^{3} \lambda_i v^{(i)} \otimes v^{(i)}.$$

By (6.14) and Hölder inequality, we have

$$(6.26)\qquad |a| \leq |\left(I_\Sigma(G_\Sigma)\right)^{-1}| \left(I_\Sigma^0(G_\Sigma)\right)^{\frac{1}{2}} \|v\|_{L^2(\Sigma,\mathbb{R}^n)}.$$

By (6.15) we have

$$(6.27)\qquad |I_\Sigma^0(G_\Sigma)| \leq C\rho_0^4,$$

where $C > 0$ only depends on α, M_0 and M_1.

By the definition of $I_\Sigma(G_\Sigma)$ we have

$$(6.28)\quad \lambda_1 = I_\Sigma(G_\Sigma)v^{(1)} \cdot v^{(1)} = \int_\Sigma |x - G_\Sigma|^2 \sin^2 \varphi(x - G_\Sigma, v^{(1)}) =$$
$$= \int_\Sigma |(x - G_\Sigma) \times v^{(1)}|^2 \equiv I_\Sigma(G_\Sigma, v^{(1)}),$$

where $I_\Sigma(G_\Sigma, v^{(1)})$ is the *moment of inertia* of Σ with respect to the straight line $v^{(1)}$ passing from G_Σ. Therefore, as $\left(I_\Sigma(G_\Sigma)\right)^{-1} = \sum_{i=1}^3 \lambda_i^{-1} v^{(i)} \otimes v^{(i)}$, to control $|\left(I_\Sigma(G_\Sigma)\right)^{-1}|$ it is enough to prove that

$$(6.29)\qquad I_\Sigma(G_\Sigma, v^{(1)}) \geq C\rho_0^4,$$

with a constant $C > 0$ only depending on M_0.

By our assumption (2.17) on the point $P_0 \in \Sigma$, we have

$$(6.30)\qquad I_\Sigma(G_\Sigma, v^{(1)}) \geq I_{\Sigma^*}(G_\Sigma, v^{(1)}),$$

where

$$(6.31)\qquad \Sigma^* = B_{\rho_0}(P_0) \cap \partial\Omega \subset \Sigma.$$

Let G_{Σ^*} be the centre of mass of Σ^*. We have

$$(6.32)\quad I_{\Sigma^*}(G_\Sigma, v^{(1)}) = I_{\Sigma^*}(G_{\Sigma^*}, v^{(1)}) + |G_\Sigma - G_{\Sigma^*}|^2|\Sigma^*| \geq I_{\Sigma^*}(G_{\Sigma^*}, v^{(1)}),$$

and therefore it remains to estimate $I_{\Sigma^*}(G_{\Sigma^*}, v^{(1)})$ from below.

Since G_{Σ^*} is an internal point of the smallest closed convex set containing Σ^*, there exist at least a point Q belonging to the intersection between the straight line passing through G_{Σ^*} with direction $v^{(1)}$ and Σ^*.

By the local representation of $\partial\Omega$ near P_0, let (x', x_3), with $x' = (x_1', x_2') \in \mathbb{R}^2$, be the coordinate system with origin in P_0 and let Q' be the projection of Q on the plane x'. We can distinguish two situations, namely: i) $|Q'-P_0| < \frac{\rho_0}{2}\cos\beta$ and ii) $|Q'-P_0| \geq \frac{\rho_0}{2}\cos\beta$, where $\tan\beta = M_0$.

In case i),

$$U^* = \left\{P \equiv (P', \psi(P')) \ s.t. \ |P'-Q'| < \frac{\rho_0}{2}\cos\beta\right\} \subset \Sigma^*, \tag{6.33}$$

and we have

$$I_{\Sigma^*}(G_{\Sigma^*}, v^{(1)}) \geq I_{U^*}(G_{\Sigma^*}, v^{(1)}). \tag{6.34}$$

To compute the integral in the right hand side of (6.34) we use the local representation of Σ near Q. In particular, it is not restrictive to choose the coordinate system (x_1', x_2') of the tangent plane to Σ at Q such that the component of $v^{(1)}$ along x_1' vanishes. Therefore, by denoting with $(y_1', y_2', \psi(y_1', y_2')) \equiv (y', \psi(y'))$ the coordinates of a generic point $P \in U^*$ and by $(0, v_2^{(1)}, v_3^{(1)})$ the components of $v^{(1)}$ in the reference system (x_1', x_2', x_3) centered in Q, with $|v^{(1)}|^2 = (v_2^{(1)})^2 + (v_3^{(1)})^2 = 1$, we have

$$\begin{aligned} I_{U^*}(G_{\Sigma^*}, v^{(1)}) &\geq \\ \geq \int_{|y'|<\frac{\rho_0}{2}\cos^2\beta} &|(y', \psi(y')) \times (0, v_2^{(1)}, v_3^{(1)})|^2 \sqrt{1+|\nabla_{y'}\psi|^2}\, dy' \geq \\ \geq \int_{|y'|<\frac{\rho_0}{2}\cos^2\beta} &\left((y_2' v_3^{(1)} - v_2^{(1)}\psi(y'))^2 + (y_1')^2\right) dy' \geq \\ &\geq \int_{|y'|<\frac{\rho_0}{2}\cos^2\beta} (y_1')^2 \, dy' = \frac{\pi}{4}\left(\frac{\rho_0}{2}\right)^4 \cos^8\beta. \end{aligned} \tag{6.35}$$

Conversely, if condition ii) holds, let us denote by S' the point belonging to the segment joining P_0 to Q' and such that $|Q'-S'| = \frac{\rho_0}{4}\cos\beta$. Then,

$$V^* = \left\{P \equiv (P', \psi(P')) \ s.t. \ |P'-S'| < \frac{\rho_0}{4}\cos\beta\right\} \subset \Sigma^* \tag{6.36}$$

and we have

$$I_{\Sigma^*}(G_{\Sigma^*}, v^{(1)}) \geq I_{V^*}(G_{\Sigma^*}, v^{(1)}). \tag{6.37}$$

Now, by calculations similar to those of case i) we have

$$I_{V^*}(G_{\Sigma^*}, v^{(1)}) \geq C\rho_0^4, \tag{6.38}$$

where $C > 0$ only depends on M_0.

By (6.35) and (6.38), the inequality (6.29) is proved.

By (6.26), (6.27) and (6.29) we have

$$|a| \leq \frac{C}{\rho_0^2}\|v\|_{L^2(\Sigma,\mathbb{R}^n)}, \tag{6.39}$$

where $C > 0$ only depends on α, M_0 and M_1.

By our regularity assumptions on the boundary we have

$$|\Sigma| \geq |B_{\rho_0}(P_0) \cap \partial\Omega| \geq \pi(\rho_0\cos\beta)^2 \tag{6.40}$$

and, therefore, by (6.13), (6.40) and Hölder inequality we have

$$|c^*| \leq \frac{C}{\rho_0}\|v\|_{L^2(\Sigma,\mathbb{R}^n)}, \tag{6.41}$$

where $C > 0$ only depends on α, M_0 and M_1.

Finally, by estimates (6.39) and (6.41) for a and c^*, respectively, by (6.15), by a trace inequality and by the global estimate (6.5), we have, for any $x \in \overline{\Omega}$,

$$|\overline{r}(x)| \leq \frac{C}{\rho_0}\|v\|_{L^2(\Sigma,\mathbb{R}^n)} + \frac{C}{\rho_0^2}\|v\|_{L^2(\Sigma,\mathbb{R}^n)}\rho_0 \leq C\|\varphi\|_{H^{-\frac{1}{2}}(\partial\Omega,\mathbb{R}^n)}, \tag{6.42}$$

where $C > 0$ only depends on α_0, β_0, M, α, M_0 and M_1. □

LEMMA 6.3. *Let Ω be a domain satisfying (2.12) and (2.18). Let D_i, $i = 1, 2$, be two connected open subsets of Ω satisfying (2.14), (2.16) and (2.19) for $D = D_i$, $i = 1, 2$. One can construct a family of regularized domains $\tilde{D}_i^h \subset \Omega$, for any h, $0 < h \leq a\rho_0$, having boundary of class C^1 with constants $\tilde{\rho}_0$ and $\tilde{M}_0$, such that*

$$D_i \subset \tilde{D}_i^{h_1} \subset \tilde{D}_i^{h_2}, \qquad 0 < h_1 \leq h_2, \tag{6.43}$$

$$\gamma_0 h \leq \operatorname{dist}(x, \partial D_i) \leq \gamma_1 h, \qquad \text{for every } x \in \partial\tilde{D}_i^h, \tag{6.44}$$

$$|\tilde{D}_i^h \setminus D_i| \leq \gamma_2 M_1 \rho_0^{n-1} h, \tag{6.45}$$

$$|\partial\tilde{D}_i^h|_{n-1} \leq \gamma_3 M_1 \rho_0^{n-1}, \tag{6.46}$$

for every $x \in \partial\tilde{D}_i^h$, there exists $y \in \partial D_i$ such that

$$|y - x| = \operatorname{dist}(x, \partial D_i), \quad |\nu(x) - \nu(y)| \leq \gamma_4 \frac{h^\alpha}{\rho_0^\alpha}, \tag{6.47}$$

where $\nu(x)$, $\nu(y)$ denote the outer unit normals to $\tilde{D}_i^h$ at x and to D_i at y respectively, and a, γ_j, $j = 0, 1, ..., 4$, and the ratios $\frac{\tilde{\rho}_0}{\rho_0}$ and $\frac{\tilde{M}_0}{M_0}$ only depend on M_0 and α. Here $|\cdot|_{n-1}$ denotes the $(n-1)$-dimensional measure.

PROOF. For the proof, based on the regularized distance introduced by Lieberman [**34**], one can argue similarly to [[**4**], Lemma 5.3]. □

In order to derive the stability estimates (4.2)– (4.4) concerning the continuation from Cauchy data, we first need to dominate the H^1 norm of $(u_1 - u_2)$ in a neighborhood of the point $P_0 \in \Sigma$ appearing in (2.17) in terms of the L^2 norm of $(u_1 - u_2)$ on Σ.

According to (2.16) and (2.20), there exists a cartesian coordinate system under which $P_0 = 0$ and

$$\Omega \cap B_{\rho_0}(0) = \{x \in B_{\rho_0}(0) \text{ s.t. } x_n > \psi(x')\} \subset \Omega_i, \quad i = 1, 2,$$

where ψ is a $C^{2,\alpha}$ function on $B'_{\rho_0}(0) \subset \mathbb{R}^{n-1}$ satisfying

$$\psi(0) = |\nabla\psi(0)| = 0$$

and

$$\|\psi\|_{C^{2,\alpha}\left(B'_{\rho_0}(0)\right)} \leq M_0\rho_0.$$

Let

$$\rho_{00} = \frac{\rho_0}{\sqrt{1 + M_0^2}}, \tag{6.48}$$

$$\Sigma_0 = \{(x', x_n) \text{ s.t. } |x'| < \rho_{00}, x_n = \psi(x')\}.$$

By (2.17) and by the definition of ρ_{00}, we have $\Sigma_0 \subset \Sigma$.

The proofs of Proposition 4.2 and Proposition 4.3 will be mainly based on a stability estimate for the solution of the Cauchy problem for the Lamé system with homogeneous Neumann data (see Proposition 6.4 below) and on a three spheres inequality for solutions to the Lamé system (see Proposition 6.5 below).

PROPOSITION 6.4. *Let Ω be a bounded domain in $\mathbb{R}^n$ and let $\Sigma \subset \partial\Omega$ be of class $C^{2,\alpha}$, with constants ρ_0, M_0. Let $u \in H^1(\Omega, \mathbb{R}^n)$ be the weak solution to the Cauchy problem*

$$\begin{cases} \operatorname{div}(\mathbb{C}\nabla u) = 0, & \text{in } \Omega, \\ (\mathbb{C}\nabla u)\nu = 0, & \text{on } \Sigma, \\ u = g & \text{on } \Sigma, \end{cases} \tag{6.49}$$

where $\mathbb{C} \in C^{1,1}(\overline{\Omega}, \mathcal{L}(\mathbb{M}^n, \mathbb{M}^n))$ satisfies (2.24)–(2.26) and $g \in H^{\frac{1}{2}}(\Sigma, \mathbb{R}^n)$. Let $P_0 \in \Sigma$ satisfy (2.17) and let $P^* = P_0 + \frac{\rho_{00}}{4}\nu$, where ν is the unit outer normal to Ω at P_0. We have

$$\|u\|_{L^\infty(\Omega\cap B_{\frac{3}{8}\rho_{00}}(P^*),\mathbb{R}^n)} + \rho_0\|\nabla u\|_{L^\infty(\Omega\cap B_{\frac{3}{8}\rho_{00}}(P^*),\mathbb{M}^n)} \leq \leq \frac{C}{\rho_0^{\frac{n}{2}}}\|u\|^{1-\tau}_{H^1(\Omega,\mathbb{R}^n)}(\rho_0^{\frac{1}{2}}\|g\|_{L^2(\Sigma,\mathbb{R}^n)})^\tau, \tag{6.50}$$

where $C > 0$ and τ, $0 < \tau < 1$, only depend on α_0, β_0, M, α and M_0.

PROOF OF PROPOSITION 6.4. The proof of this Proposition was obtained in [**36**, Theorem 5.3]. □

PROPOSITION 6.5 (Three Spheres Inequality). *Let Ω be a domain in $\mathbb{R}^n$, $n = 2$ or $n = 3$, and let the elasticity tensor $\mathbb{C}$ satisfy (2.24)–(2.26). Let $u \in H^1(\Omega, \mathbb{R}^n)$ be a solution to the Lamé system (2.32). There exists ϑ^*, $0 < \vartheta^* \leq 1$, only depending on α_0, β_0 and M, such that for every $\rho_1, \rho_2, \rho_3, \tilde{\rho}$, $0 < \rho_1 < \rho_2 < \rho_3 \leq \vartheta^*\tilde{\rho}$, and for every $x \in \Omega_{\tilde{\rho}}$ we have*

$$\int_{B_{\rho_2}(x)} |u|^2 \leq C\left(\int_{B_{\rho_1}(x)} |u|^2\right)^\delta \left(\int_{B_{\rho_3}(x)} |u|^2\right)^{1-\delta}, \tag{6.51}$$

where $C > 0$ and δ, $0 < \delta < 1$, only depend on α_0, β_0, M, $\frac{\rho_2}{\rho_3}$ and $\frac{\rho_1}{\rho_3}$.

PROOF OF PROPOSITION 6.5. A proof of this Proposition was obtained in [**5**], see also [**6**, Corollary 3.3]. □

PROOF OF PROPOSITION 4.2. The proof of this Proposition is rather technical even in the simpler case when the infinitesimal rigid displacement $\overline{r}$ appearing in (2.38) and minimizing the distance in $L^2(\Sigma)$ of $u_1 - u_2$ from $\mathcal{R}$, vanishes. Moreover, significant new difficulties occur in the general case $\overline{r} \neq 0$. Therefore, for a better comprehension of the arguments involved, we shall divide the proof into two main steps. In the first one we shall treat the case $\overline{r} = 0$ for both dimension $n = 2$ and $n = 3$. In the second step we shall consider the general case and, for the sake of simplicity, we shall focus our attention on the two-dimensional case. The three dimensional case when $\overline{r} \neq 0$ needs some more technical details and shall be discussed in Chapter 7.

In the sequel we shall prove (4.2), (4.4), the proof of (4.3), (4.4) being analogous.

Step 1: $\overline{r}=0$.

It is not restrictive to assume, for this proof, that $\epsilon \leq \rho_0^{\frac{3-n}{2}} \|\varphi\|_{H^{-\frac{1}{2}}(\partial\Omega,\mathbb{R}^n)}\tilde{\mu}$, where $\tilde{\mu}$, $0<\tilde{\mu}<e^{-1}$, is a constant, only depending on α_0, β_0, M, α, M_0 and M_1, which will be chosen later on. In fact, otherwise, (4.2)–(4.4) become trivial.

Let $\theta = \min\{a, \frac{7}{8\gamma_1}, \frac{\rho_{00}}{2\gamma_0\sqrt{1+M_0^2}}\}$, where a, γ_0, γ_1 have been introduced in Lemma 6.3. We have that θ only depends on α and M_0. Let $\overline{\rho}=\theta\rho_0$ and let $\rho\leq\overline{\rho}$.

Let us denote by $\tilde{V}_\rho$ the connected component of $\overline{\Omega}\setminus(\tilde{D}_1^\rho\cup\tilde{D}_2^\rho)$ which contains $\partial\Omega$. We have

$$D_2\setminus\overline{D_1}\subset\Omega_1\setminus\overline{G}\subset\left((\tilde{D}_1^\rho\setminus\overline{D_1})\setminus\overline{G}\right)\cup\left((\Omega\setminus\tilde{V}_\rho)\setminus\overline{\tilde{D}_1^\rho}\right),$$

$$\partial\left((\Omega\setminus\tilde{V}_\rho)\setminus\overline{\tilde{D}_1^\rho}\right)=\tilde{\Gamma}_1^\rho\cup\tilde{\Gamma}_2^\rho,$$

where $\tilde{\Gamma}_2^\rho=\partial\tilde{D}_2^\rho\cap\partial\tilde{V}_\rho$ and $\tilde{\Gamma}_1^\rho\subset\partial\tilde{D}_1^\rho$. We have

$$\int_{D_2\setminus D_1}|\widehat{\nabla}u_1|^2\leq\int_{\Omega_1\setminus G}|\widehat{\nabla}u_1|^2\leq\int_{(\tilde{D}_1^\rho\setminus\overline{D_1})\setminus\overline{G}}|\widehat{\nabla}u_1|^2+\int_{(\Omega\setminus\tilde{V}_\rho)\setminus\overline{\tilde{D}_1^\rho}}|\widehat{\nabla}u_1|^2. \tag{6.52}$$

By (6.2) and (6.45) we have

$$\int_{(\tilde{D}_1^\rho\setminus\overline{D_1})\setminus\overline{G}}|\widehat{\nabla}u_1|^2\leq C\|\varphi\|^2_{H^{-\frac{1}{2}}(\partial\Omega,\mathbb{R}^n)}\rho\leq C\rho_0\|\varphi\|^2_{H^{-\frac{1}{2}}(\partial\Omega,\mathbb{R}^n)}\frac{\rho}{\rho_0}, \tag{6.53}$$

with $C>0$ only depending on α_0, β_0, M, α, M_0 and M_1. By applying the divergence theorem we have

$$\int_{(\Omega\setminus\tilde{V}_\rho)\setminus\overline{\tilde{D}_1^\rho}}|\widehat{\nabla}u_1|^2\leq\xi_0^{-1}\left(\int_{\tilde{\Gamma}_1^\rho}(\mathbb{C}\nabla u_1)\nu\cdot u_1+\int_{\tilde{\Gamma}_2^\rho}(\mathbb{C}\nabla u_1)\nu\cdot u_1\right). \tag{6.54}$$

Let $x\in\tilde{\Gamma}_1^\rho$. By (6.44), $\operatorname{dist}(x,\partial D_1)\leq\gamma_1\rho$. Hence, there exists $y\in\partial D_1$ such that $|y-x|=\operatorname{dist}(x,\partial D_1)\leq\gamma_1\rho$. Since $u_1(y)=0$, from (6.2) and (6.44) we have

$$|u_1(x)|\leq\frac{C}{\rho_0^{\frac{n-3}{2}}}\|\varphi\|_{H^{-\frac{1}{2}}(\partial\Omega,\mathbb{R}^n)}\frac{\rho}{\rho_0}, \tag{6.55}$$

where $C>0$ only depends on α_0, β_0, M, α, M_0 and M_1.

Given $x\in\tilde{\Gamma}_2^\rho$, one can prove similarly that there exists $y\in\partial D_2$ such that $|y-x|=\operatorname{dist}(x,\partial D_2)\leq\gamma_1\rho$. Since $u_2(y)=0$, we have

$$|u_1(x)|\leq C\left(\frac{\|\varphi\|_{H^{-\frac{1}{2}}(\partial\Omega,\mathbb{R}^n)}}{\rho_0^{\frac{n-3}{2}}}\frac{\rho}{\rho_0}+|w(x)|\right), \tag{6.56}$$

where $C>0$ only depends on α_0, β_0, M, α, M_0 and M_1, and

$$w=u_1-u_2. \tag{6.57}$$

From (6.2), (6.46), (6.52)-(6.56), we have

$$\int_{\Omega_1\setminus G}|\widehat{\nabla}u_1|^2\leq C\rho_0\left(\|\varphi\|^2_{H^{-\frac{1}{2}}(\partial\Omega,\mathbb{R}^n)}\frac{\rho}{\rho_0}+\rho_0^{\frac{n-3}{2}}\|\varphi\|_{H^{-\frac{1}{2}}(\partial\Omega,\mathbb{R}^n)}\max_{\partial\tilde{V}_\rho\setminus\partial\Omega}|w|\right) \tag{6.58}$$

where C only depends on α_0, β_0, M, α, M_0 and M_1.

In order to estimate $\max_{\partial\tilde{V}_\rho\setminus\partial\Omega}|w|$, we shall make use of the stability estimate for the Cauchy problem stated in Theorem 6.4 and we shall perform a propagation of smallness argument, based on an iterated application of the three spheres inequality (6.51).

Let

$$z_0 = P_0 - \frac{\rho_1}{16}\nu, \tag{6.59}$$

$$\rho^* = \frac{\rho_0}{16(1+M_0^2)}, \tag{6.60}$$

where ν denotes the outer unit normal to Ω at P_0.

In order to develop the above mentioned arguments, let us consider the set $\tilde{V}_\rho \cap \overline{\Omega_{\frac{\rho^*}{2}}}$. By the choice of $\overline{\rho}$, this set is connected and contains z_0. Let x be any point in $\partial\tilde{V}_\rho \setminus \partial\Omega$ and let γ be a path in $\tilde{V}_\rho \cap \overline{\Omega_{\frac{\rho^*}{2}}}$ joining x to z_0. Let us define $\{x_i\}$, $i = 1, ..., s$, as follows: $x_1 = z_0$, $x_{i+1} = \gamma(t_i)$, where

$$t_i = \max\{t \text{ s.t. } |\gamma(t) - x_i| = \frac{\gamma_0\rho\theta^*}{2}\}, \quad \text{if} |x_i - x| > \frac{\gamma_0\rho\theta^*}{2},$$

otherwise let $i = s$ and stop the process. By construction, the balls $B_{\frac{\gamma_0\rho\theta^*}{4}}(x_i)$ are pairwise disjoint, $|x_{i+1} - x_i| = \frac{\gamma_0\rho\theta^*}{2}$, for $i = 1, ..., s-1$, $|x_s - x| \le \frac{\gamma_0\rho\theta^*}{2}$. Hence we have $s \le S\left(\frac{\rho_0}{\rho}\right)^n$, with S only depending on α_0, β_0, M, α, M_0 and M_1.

An iterated application of the three spheres inequality (6.51) for w with radii $\rho_1 = \frac{\gamma_0\rho\theta^*}{4}$, $\rho_2 = \frac{3\gamma_0\rho\theta^*}{4}$, $\rho_3 = \gamma_0\rho\theta^*$, gives that for every ρ, $0 < \rho \le \overline{\rho}$,

$$\int_{B_{\frac{\gamma_0\rho\theta^*}{4}}(x)} |w|^2 \le C\left(\int_G |w|^2\right)^{1-\delta^s} \left(\int_{B_{\frac{\gamma_0\rho\theta^*}{4}}(z_0)} |w|^2\right)^{\delta^s}, \tag{6.61}$$

where δ, $0 < \delta < 1$, $C \ge 1$, only depend on α_0, β_0 and M. From now on, let us denote

$$\tilde{\epsilon} = \frac{\epsilon}{\rho_0^{\frac{3-n}{2}} \|\varphi\|_{H^{-\frac{1}{2}}(\partial\Omega,\mathbb{R}^n)}}. \tag{6.62}$$

By the choice of $\overline{\rho}$, $B_{\frac{\gamma_0\rho\theta^*}{4}}(z_0) \subset B_{\rho^*}(z_0) \subset G \cap B_{\frac{3}{8}\rho_1}(P^*)$ and we can apply Theorem 6.4 to estimate the right hand side of (6.61). By (6.5), (6.50), (2.38) and (6.61) we obtain

$$\int_{B_{\frac{\gamma_0\rho\theta^*}{4}}(x)} |w|^2 \le C\rho_0^3 \|\varphi\|^2_{H^{-\frac{1}{2}}(\partial\Omega,\mathbb{R}^n)} \tilde{\epsilon}^{2\tau\delta^s}, \tag{6.63}$$

where τ, $0 < \tau < 1$, and $C \ge 1$ depend on α_0, β_0, M, α, M_0 and M_1 only. At this stage, let us recall the following interpolation inequality

$$\|v\|_{L^\infty(B_t)} \le C\left(\left(\int_{B_t} |v|^2\right)^{\frac{1}{n+2}} |\nabla v|^{\frac{n}{n+2}}_{L^\infty(B_t)} + \frac{1}{t^{n/2}}\left(\int_{B_t} |v|^2\right)^{1/2}\right), \tag{6.64}$$

which holds for any function v defined in the ball $B_t \subset \mathbb{R}^n$. By applying (6.64) to w in $B_{\frac{\gamma_0\rho\theta^*}{4}}(x)$ and by using (6.2) and (6.63), we obtain

$$\|w\|_{L^\infty(B_{\frac{\gamma_0\rho\theta^*}{4}}(x),\mathbb{R}^n)} \le \frac{C}{\rho_0^{\frac{n-3}{2}}}\left(\frac{\rho_0}{\rho}\right)^{n/2} \|\varphi\|_{H^{-\frac{1}{2}}(\partial\Omega,\mathbb{R}^n)} \tilde{\epsilon}^{\gamma\delta^s}, \tag{6.65}$$

where $\gamma = \frac{2\tau}{n+2}$, $0 < \gamma < 1$, and C depends on α_0, β_0, M, M_0 and M_1 only. From (6.58) and (6.65) we have that for any $\rho \leq \bar{\rho}$

$$\int_{\Omega_1 \setminus G} |\widehat{\nabla} u_1|^2 \leq C\rho_0 \|\varphi\|^2_{H^{-\frac{1}{2}}(\partial\Omega,\mathbb{R}^n)} \left(\frac{\rho}{\rho_0} + \left(\frac{\rho_0}{\rho}\right)^{n/2} \tilde{\epsilon}^{\gamma\delta^s}\right), \tag{6.66}$$

with C only depending on α_0, β_0, M, α, M_0 and M_1.

Let us set $\bar{\mu} = \exp\left\{-\frac{1}{\gamma}\exp\left(\frac{2S|\log\delta|}{\theta^n}\right)\right\}$, $\tilde{\mu} = \min\{\bar{\mu}, \exp(-\gamma^{-2})\}$. We have that $\tilde{\mu} < e^{-1}$ and it depends on α_0, β_0, M, α, M_0 and M_1 only. Let $\tilde{\epsilon} \leq \tilde{\mu}$ and let

$$\rho(\tilde{\epsilon}) = \rho_0 \left(\frac{2S|\log\delta|}{\log|\log\tilde{\epsilon}^\gamma|}\right)^{1/n}. \tag{6.67}$$

Since $\rho(\tilde{\epsilon})$ is increasing in $(0, e^{-1})$ and since $\rho(\tilde{\mu}) \leq \rho(\bar{\mu}) = \rho_0\theta = \bar{\rho}$, we can apply inequality (6.66) with $\rho = \rho(\tilde{\epsilon})$, obtaining

$$\int_{\Omega_1 \setminus G} |\widehat{\nabla} u_1|^2 \leq C\rho_0 \|\varphi\|^2_{H^{-\frac{1}{2}}(\partial\Omega,\mathbb{R}^n)} \left(\log|\log\tilde{\epsilon}^\gamma|\right)^{-1/n}, \tag{6.68}$$

where C only depends on α_0, β_0, M, α, M_0 and M_1.

Since $\tilde{\epsilon} \leq \exp(-\gamma^{-2})$, we also have that $\log\gamma \geq -\frac{1}{2}\log|\log\tilde{\epsilon}|$, so that

$$\log|\log\tilde{\epsilon}^\gamma| \geq \frac{1}{2}\log|\log\tilde{\epsilon}|. \tag{6.69}$$

From (6.68) and (6.69) we have

$$\int_{D_2 \setminus D_1} |\widehat{\nabla} u_1|^2 \leq \rho_0 \|\varphi\|^2_{H^{-\frac{1}{2}}(\partial\Omega,\mathbb{R}^n)} \omega^*(\tilde{\epsilon}), \tag{6.70}$$

with

$$\omega^*(t) = C\left(\log|\log t|\right)^{-\frac{1}{n}} \quad \text{for every } t < e^{-1}, \tag{6.71}$$

where $C > 0$ is a constant only depending on α_0, β_0, M, α, M_0 and M_1.

Step 2: $\overline{r} \neq 0$ and $n = 2$.

In the sequel we shall use the notation introduced in Step 1. Let us denote

$$\tilde{\rho} = \rho(\tilde{\epsilon}), \tag{6.72}$$

where $\rho(\tilde{\epsilon})$ is given by (6.67).

We can distinguish the following three cases:

I) $\partial\tilde{D}_1^{\tilde{\rho}} \cap \tilde{\Gamma}_2^{\tilde{\rho}} = \emptyset$;

II) there exist at least two points z_1 and z_2, $z_i \in \partial\tilde{D}_1^{\tilde{\rho}} \cap \tilde{\Gamma}_2^{\tilde{\rho}}$, $i = 1, 2$, satisfying

$$|z_1 - z_2| \geq \rho_0 \left(\log|\log\tilde{\epsilon}|\right)^{-\frac{1}{2n}}; \tag{6.73}$$

III) $\operatorname{diam}\left(\partial\tilde{D}_1^{\tilde{\rho}} \cap \tilde{\Gamma}_2^{\tilde{\rho}}\right) \leq \rho_0 \left(\log|\log\tilde{\epsilon}|\right)^{-\frac{1}{2n}}$.

When I) holds, there are three possible subcases:

Ia) $\tilde{D}_1^{\tilde{\rho}} \cap \tilde{D}_2^{\tilde{\rho}} = \emptyset$,

Ib) $\tilde{D}_1^{\tilde{\rho}} \subset \tilde{D}_2^{\tilde{\rho}}$,

Ic) $\tilde{D}_2^{\tilde{\rho}} \subset \tilde{D}_1^{\tilde{\rho}}$.

In case Ia) we have that $(\Omega \setminus \tilde{V}_{\tilde{\rho}}) \setminus \tilde{D}_1^{\tilde{\rho}} = \tilde{D}_2^{\tilde{\rho}}$ and, therefore, it follows that $\partial\left((\Omega \setminus \tilde{V}_{\tilde{\rho}}) \setminus \tilde{D}_1^{\tilde{\rho}}\right) = \partial\tilde{D}_2^{\tilde{\rho}}$, whereas in case Ib) we have that $(\Omega \setminus \tilde{V}_{\tilde{\rho}}) \setminus \tilde{D}_1^{\tilde{\rho}} = \tilde{D}_2^{\tilde{\rho}} \setminus \tilde{D}_1^{\tilde{\rho}}$.

For both cases Ia) and Ib), by applying the divergence theorem to u_1 in $\tilde{D}_2^{\tilde{\rho}}$ and in $\tilde{D}_2^{\tilde{\rho}} \setminus D_1$ respectively, we have

$$\int_{\partial \tilde{D}_2^{\tilde{\rho}}} (\mathbb{C}\nabla u_1)\nu \cdot r = 0, \quad \text{for every } r \in \mathcal{R}. \tag{6.74}$$

Let us set

$$w = u_1 - u_2 - \overline{r}. \tag{6.75}$$

By applying the estimates of continuation from Cauchy data obtained in the above step (see (6.65)) to w, we have

$$\|w\|_{L^\infty(\partial \tilde{V}_{\tilde{\rho}} \setminus \partial\Omega, \mathbb{R}^n)} \leq \frac{C}{\rho_0^{\frac{n-3}{2}}} \left(\frac{\rho_0}{\tilde{\rho}}\right)^{n/2} \|\varphi\|_{H^{-\frac{1}{2}}(\partial\Omega,\mathbb{R}^n)} \tilde{\epsilon}^{\gamma\delta^s}, \tag{6.76}$$

where $C > 0$ only depends on α_0, β_0, M, M_0, M_1 and $\gamma = \frac{2\tau}{n+2}$, $0 < \gamma < 1$.

By recalling that $u_i = 0$ on ∂D_i, $i = 1, 2$, and by (6.74), (6.2), (6.46), (6.76) and (6.72) we have, for both cases Ia) and Ib),

$$\begin{aligned} \int_{D_2 \setminus D_1} |\widehat{\nabla} u_1|^2 &\leq \int_{\tilde{D}_2^{\tilde{\rho}} \setminus D_1} |\widehat{\nabla} u_1|^2 \leq \xi_0^{-1} \int_{\partial \tilde{D}_2^{\tilde{\rho}}} (\mathbb{C}\nabla u_1)\nu \cdot u_1 = \\ &= \xi_0^{-1} \left(\int_{\partial \tilde{D}_2^{\tilde{\rho}}} (\mathbb{C}\nabla u_1)\nu \cdot u_2 + \int_{\partial \tilde{D}_2^{\tilde{\rho}}} (\mathbb{C}\nabla u_1)\nu \cdot w \right) \leq \\ &\leq C\rho_0 \|\varphi\|^2_{H^{-\frac{1}{2}}(\partial\Omega,\mathbb{R}^n)} \left(\frac{\tilde{\rho}}{\rho_0} + \left(\frac{\rho_0}{\tilde{\rho}}\right)^{n/2} \tilde{\epsilon}^{\gamma\delta^s} \right) = \rho_0 \|\varphi\|^2_{H^{-\frac{1}{2}}(\partial\Omega,\mathbb{R}^n)} \omega^*(\tilde{\epsilon}). \end{aligned} \tag{6.77}$$

In case Ic), by using (6.45), we have

$$|D_2 \setminus D_1| \leq |\tilde{D}_1^{\tilde{\rho}} \setminus D_1| \leq C\rho_0^{n-1}\tilde{\rho}, \tag{6.78}$$

with C only depending on α, M_0, M_1. By the above inequality and by (6.2), we have

$$\int_{D_2 \setminus D_1} |\widehat{\nabla} u_1|^2 \leq C\rho_0 \|\varphi\|^2_{H^{-\frac{1}{2}}(\partial\Omega,\mathbb{R}^n)} \frac{\tilde{\rho}}{\rho_0}, \tag{6.79}$$

with C only depending on α_0, β_0, M, α, M_0, M_1, so that trivially (6.70) and (6.71) follow.

Let us consider now case II). In view of the above arguments, it is clear from (6.76) and by the choice of $\tilde{\rho}$ that

$$\|w\|_{L^\infty(\partial \tilde{V}_{\tilde{\rho}} \setminus \partial\Omega, \mathbb{R}^n)} \leq \frac{C}{\rho_0^{\frac{n-3}{2}}} \|\varphi\|_{H^{-\frac{1}{2}}(\partial\Omega,\mathbb{R}^n)} \omega^*(\tilde{\epsilon}), \tag{6.80}$$

where $C > 0$ only depends on α_0, β_0, M, α, M_0 and M_1. Therefore, we have

$$|w(z_j)| \leq \frac{C}{\rho_0^{\frac{n-3}{2}}} \|\varphi\|_{H^{-\frac{1}{2}}(\partial\Omega,\mathbb{R}^n)} \omega^*(\tilde{\epsilon}), \quad \text{j=1,2}, \tag{6.81}$$

where $C > 0$ only depends on α_0, β_0, M, α, M_0 and M_1. Moreover, recalling the homogeneous Dirichlet condition for u_i on ∂D_i, $i = 1, 2$, and by (6.2) we have

$$|u_i(z_j)| \leq \frac{C}{\rho_0^{\frac{n-3}{2}}} \|\varphi\|_{H^{-\frac{1}{2}}(\partial\Omega,\mathbb{R}^n)} \frac{\tilde{\rho}}{\rho_0}, \quad \text{i, j=1,2}, \tag{6.82}$$

where $C > 0$ only depends on α_0, β_0, M, α, M_0 and M_1.

From (6.81) and (6.82) it follows that

$$|\overline{r}(z_j)| \leq \frac{C}{\rho_0^{\frac{n-3}{2}}} \|\varphi\|_{H^{-\frac{1}{2}}(\partial\Omega,\mathbb{R}^n)} (\log|\log\tilde{\epsilon}|)^{-\frac{1}{n}}, \quad \text{j=1,2,} \tag{6.83}$$

where $C > 0$ only depends on α_0, β_0, M, α, M_0 and M_1.

By embedding $\mathbb{R}^2$ in $\mathbb{R}^3$ and with the obvious notation, $\overline{r}(x) = c + a_3 e_3 \times x$, where $c = c_1e_1 + c_2e_2$, with c_1, c_2 and a_3 real constants.

By (6.83) and (6.73) we have

$$|a_3| (\log|\log\tilde{\epsilon}|)^{-\frac{1}{2n}} \rho_0 \leq |a_3||z_1 - z_2| = \tag{6.84}$$

$$= |\overline{r}(z_1) - \overline{r}(z_2)| \leq \frac{C}{\rho_0^{\frac{n-3}{2}}} \|\varphi\|_{H^{-\frac{1}{2}}(\partial\Omega,\mathbb{R}^n)} (\log|\log\tilde{\epsilon}|)^{-\frac{1}{n}},$$

so that

$$|a_3| \leq \frac{C}{\rho_0^{\frac{n-1}{2}}} \|\varphi\|_{H^{-\frac{1}{2}}(\partial\Omega,\mathbb{R}^n)} (\log|\log\tilde{\epsilon}|)^{-\frac{1}{2n}}, \tag{6.85}$$

where $C > 0$ only depends on α_0, β_0, M, α, M_0 and M_1.

By (6.83) and (6.85) we have

$$|c| \leq \frac{C}{\rho_0^{\frac{n-3}{2}}} \|\varphi\|_{H^{-\frac{1}{2}}(\partial\Omega,\mathbb{R}^n)} (\log|\log\tilde{\epsilon}|)^{-\frac{1}{2n}}, \tag{6.86}$$

where $C > 0$ only depends on α_0, β_0, M, α, M_0 and M_1.

Finally, by (6.85) and (6.86) we have

$$\|\overline{r}\|_{L^\infty(\Omega)} \leq \frac{C}{\rho_0^{\frac{n-3}{2}}} \|\varphi\|_{H^{-\frac{1}{2}}(\partial\Omega,\mathbb{R}^n)} (\log|\log\tilde{\epsilon}|)^{-\frac{1}{2n}}, \tag{6.87}$$

where $C > 0$ only depends on α_0, β_0, M, α, M_0 and M_1.

Now, the thesis follows by repeating the arguments of Step 1 for the function $w = u_1 - u_2 - \overline{r}$, the only difference consists in the appearance of the additional term $|\overline{r}(x)|$ in the right hand side of (6.56), which can be controlled by (6.87). Therefore, we obtain

$$\int_{D_2\setminus D_1} |\widehat{\nabla} u_1|^2 \leq \rho_0 \|\varphi\|^2_{H^{-\frac{1}{2}}(\partial\Omega,\mathbb{R}^n)} \omega^*(\tilde{\epsilon})^{\frac{1}{2}}. \tag{6.88}$$

Finally, let us consider the case III). We have

$$\int_{(\Omega\setminus\tilde{V}_{\bar{\rho}})\setminus\tilde{D}_1^{\bar{\rho}}} |\widehat{\nabla} u_1|^2 \leq \xi_0^{-1} \int_{(\Omega\setminus\tilde{V}_{\bar{\rho}})\setminus\tilde{D}_1^{\bar{\rho}}} \mathbb{C}\nabla u_1 \cdot \nabla u_1 = \tag{6.89}$$

$$= \xi_0^{-1} \int_{(\Omega\setminus\tilde{V}_{\bar{\rho}})\setminus\tilde{D}_1^{\bar{\rho}}} \mathbb{C}\nabla(u_1 - \overline{r}) \cdot \nabla(u_1 - \overline{r}) =$$

$$= \xi_0^{-1} \left\{ \int_{\tilde{\Gamma}_2^{\bar{\rho}}} (\mathbb{C}\nabla u_1)\nu \cdot (u_1 - \overline{r}) + \int_{\tilde{\Gamma}_1^{\bar{\rho}}} (\mathbb{C}\nabla u_1)\nu \cdot u_1 - \int_{\tilde{\Gamma}_1^{\bar{\rho}}} (\mathbb{C}\nabla u_1)\nu \cdot \overline{r} \right\}.$$

The first addend in the right hand side of the above inequality can be estimated as usual by decomposing $u_1 - \overline{r} = w + u_2$ and by using the estimate of continuation for $w = u_1 - u_2 - \overline{r}$. Also the second integral is easily estimated repeating arguments just seen in previous steps.

Let σ be the smallest subarc of $\partial\tilde{D}_1^{\tilde{\rho}}$ containing $\partial\tilde{D_1}^{\tilde{\rho}} \cap \tilde{\Gamma}_2^{\tilde{\rho}}$, that is σ is the intersection of all the connected subsets of $\partial\tilde{D}_1^{\tilde{\rho}}$ containing $\partial\tilde{D_1}^{\tilde{\rho}} \cap \tilde{\Gamma}_2^{\tilde{\rho}}$. By our regularity assumptions and by Lemma 6.3, we have

$$\text{length}(\sigma) \leq C\rho_0(\log|\log\tilde{\epsilon}|)^{-\frac{1}{2n}}, \tag{6.90}$$

with C only depending on M_0, M_1 and α. It is evident that a path on $\partial\tilde{D}_1^{\tilde{\rho}}$ connecting a point of $\tilde{\Gamma}_1^{\tilde{\rho}}$ with a point of $\partial\tilde{D}_1^{\tilde{\rho}} \setminus \tilde{\Gamma}_1^{\tilde{\rho}}$ must intersect $\partial\tilde{D}_1^{\tilde{\rho}} \cap \tilde{\Gamma}_2^{\tilde{\rho}}$. Since $\partial\tilde{D}_1^{\tilde{\rho}} \setminus \sigma$ is connected and does not intersect $\tilde{\Gamma}_2^{\tilde{\rho}} \cap \partial\tilde{D}_1^{\tilde{\rho}}$, then it cannot intersect both $\tilde{\Gamma}_1^{\tilde{\rho}}$ and $\partial\tilde{D}_1^{\tilde{\rho}} \setminus \tilde{\Gamma}_1^{\tilde{\rho}}$. Therefore either $\tilde{\Gamma}_1^{\tilde{\rho}} \subset \sigma$ or $\tilde{\Gamma}_1^{\tilde{\rho}} \supset \partial\tilde{D}_1^{\tilde{\rho}} \setminus \sigma$. In the former case the third integral in (6.89) is easily bounded by recalling (6.2), (6.90) and Lemma 6.2. In the latter case, by applying the divergence theorem to u_1 in $\Omega \setminus \tilde{D}_1^{\tilde{\rho}}$, we obtain

$$\int_{\partial\tilde{D}_1^{\tilde{\rho}}} (\mathbb{C}\nabla u_1)\nu \cdot \overline{r} = 0. \tag{6.91}$$

Noticing that in this case $\partial\tilde{D}_1^{\tilde{\rho}} \setminus \tilde{\Gamma}_1^{\tilde{\rho}} \subset \sigma$, we have

$$-\int_{\tilde{\Gamma}_1^{\tilde{\rho}}} (\mathbb{C}\nabla u_1)\nu \cdot \overline{r} = \int_{\partial\tilde{D}_1^{\tilde{\rho}} \setminus \tilde{\Gamma}_1^{\tilde{\rho}}} (\mathbb{C}\nabla u_1)\nu \cdot \overline{r} \leq \int_\sigma |(\mathbb{C}\nabla u_1)\nu \cdot \overline{r}|, \tag{6.92}$$

so that we reduce to the previous case.

Therefore, collecting all cases, (4.2), (4.4) hold with $c_2 = \frac{1}{4}$.

We recall that the proof for the case $n = 3$ is given in Chapter 7. □

PROOF OF PROPOSITION 4.3. As in the proof of Proposition 4.2, let us consider first the case $\overline{r} = 0$. Also for this proof, it is not restrictive to assume $\epsilon \leq r_0^{\frac{3-n}{2}} \|\varphi\|_{H^{-\frac{1}{2}}(A,\mathbb{R}^n)}\tilde{\mu}$, where $\tilde{\mu}$, $0 < \tilde{\mu} < e^{-1}$, is a constant, only depending on α_0, β_0, M, α, M_0 and M_1, which will be chosen later on.

Let us prove (4.2) and (4.5), the case of (4.3) and (4.5) being analogous.

We have

$$\int_{D_2\setminus D_1} |\widehat{\nabla} u_1|^2 \leq \int_{\Omega_1\setminus G} |\widehat{\nabla} u_1|^2 \leq \xi_0^{-1} \int_{\partial(\Omega_1\setminus G)} (\mathbb{C}\nabla u_1)\nu \cdot u_1, \tag{6.93}$$

and

$$\partial(\Omega_1 \setminus G) \subset \partial D_1 \cup (\partial D_2 \cap \partial G).$$

From the boundary condition $u_i = 0$ on ∂D_i, $i = 1, 2$, and by (6.2) it follows that

$$\int_{\Omega_1\setminus G} |\widehat{\nabla} u_1|^2 \leq \xi_0^{-1} \int_{\partial D_2\cap\partial G} (\mathbb{C}\nabla u_1)\nu \cdot w \leq C\rho_0^{\frac{n-1}{2}} \|\varphi\|_{H^{-\frac{1}{2}}(\partial\Omega,\mathbb{R}^n)} \max_{\partial G} |w|, \tag{6.94}$$

where $w = u_1 - u_2$ and where C only depends on α_0, β_0, M, α, M_0 and M_1.

Let us introduce the following notation. Given $z \in \mathbb{R}^n$, $\xi \in \mathbb{R}^n$, $|\xi| = 1$, $\vartheta > 0$, we shall denote by

$$C(z, \xi, \vartheta) = \{x \in \mathbb{R}^n \text{ s. t. } \frac{(x - z) \cdot \xi}{|x - z|} > \cos\vartheta\}, \tag{6.95}$$

the open cone having vertex z, axis in the direction ξ and width 2ϑ.

By our regularity hypotheses on ∂G, it follows that for every $z \in \partial G$ there exists $\xi \in \mathbb{R}^n$, $|\xi| = 1$, such that $C(z, \xi, \vartheta_0) \cap B_{\tilde{r}_0}(z) \subset G$, where $\vartheta_0 = \arctan\frac{1}{M_0}$. Notice also that G_ρ is connected for $\rho \leq \frac{\tilde{\rho}_0}{3}$. Let us fix $z \in \partial G$ and set

$$\lambda_1 = \min\{\frac{\tilde{\rho}_0}{1 + \sin\vartheta_0}, \frac{\tilde{\rho}_0}{3\sin\vartheta_0}, \frac{\rho_0}{16(1 + M_0^2)\sin\vartheta_0}\},$$

$$\vartheta_1 = \arcsin\left(\frac{\sin\vartheta_0}{4}\right),$$
$$w_1 = z + \lambda_1\xi,$$
$$\rho_1 = \vartheta^*\lambda_1 \sin\vartheta_1.$$

where ϑ^*, $0 < \vartheta^* \leq 1$, only depending on α_0, β_0, M, has been introduced in Lemma 6.5. By construction,

$$B_{\rho_1}(w_1) \subset C(z,\xi,\vartheta_1) \cap B_{\tilde{\rho}_0}(z),$$
$$B_{\frac{4\rho_1}{\vartheta^*}}(w_1) \subset C(z,\xi,\vartheta_0) \cap B_{\tilde{\rho}_0}(z) \subset G.$$

Moreover $\frac{4\rho_1}{\vartheta^*} \leq \rho^*$, so that $B_{\frac{4\rho_1}{\vartheta^*}}(z_0) \subset G$, where z_0 and ρ^* have been defined in the proof of Proposition 4.2 by (6.59) and (6.60), respectively. Hence both w_1 and z_0 belong to $\overline{G_{\frac{4\rho_1}{\vartheta^*}}}$, which is connected since $\rho_1 \leq \frac{\tilde{\rho}_0\vartheta^*}{12}$. By an iterated application of the three spheres inequality (6.51) for w we have

$$\int_{B_{\rho_1}(w_1)} |w|^2 \leq C\left(\int_G |w|^2\right)^{1-\delta^s}\left(\int_{B_{\rho_1}(z_0)} |w|^2\right)^{\delta^s}, \tag{6.96}$$

where δ, $0 < \delta < 1$, and $C \geq 1$ only depend on α_0, β_0, M, and where $s \leq \frac{M_1\rho_0^n}{\omega_n\rho_1^n}$. Since $B_{\rho^*}(z_0) \subset G \cap B_{\frac{3}{8}\rho_{00}}(P^*)$, we can apply Proposition 6.4 to w and, by (6.5) and (2.38), we have

$$\int_{B_{\rho_1}(w_1)} |w|^2 \leq C\rho_0^3\|\varphi\|^2_{H^{-\frac{1}{2}}(\partial\Omega,\mathbb{R}^n)}\tilde{\epsilon}^{2\beta_1}, \tag{6.97}$$

where β_1, $0 < \beta_1 < 1$ and $C \geq 1$ only depend on α_0, β_0, M, α, M_0, M_1 and $\frac{\tilde{\rho}_0}{\rho_0}$ and $\tilde{\epsilon}$ is given by (6.62).

The next step consists in approaching $z \in \partial G$ by constructing a sequence of balls contained in $C(z,\xi,\vartheta_1)$ as follows. Let us define, for $k \geq 2$,

$$w_k = z + \lambda_k\xi,$$
$$\lambda_k = \chi\lambda_{k-1},$$
$$\rho_k = \chi\rho_{k-1},$$

with

$$\chi = \frac{1-\sin\vartheta_1}{1+\sin\vartheta_1}.$$

We have that

$$\rho_k = \chi^{k-1}\rho_1,$$
$$\lambda_k = \chi^{k-1}\lambda_1,$$
$$B_{\rho_{k+1}}(w_{k+1}) \subset B_{3\rho_k}(w_k)$$

and

$$B_{\frac{4}{\vartheta^*}\rho_k}(w_k) \subset C(z,\xi,\vartheta_0) \cap B_{\tilde{\rho}_0}(z) \subset G.$$

Denoting

$$d(k) = |w_k - z| - \rho_k,$$

we have

$$d(k) = \chi^{k-1}d(1),$$

with

$$d(1) = \lambda_1(1 - \vartheta^*\sin\vartheta_1).$$

For any t, $0 < t \leq d(1)$, let $k(t)$ be the smallest positive integer such that $d(k) \leq t$, that is

$$\frac{\left|\log \frac{t}{d(1)}\right|}{|\log \chi|} \leq k(t) - 1 \leq \frac{\left|\log \frac{t}{d(1)}\right|}{|\log \chi|} + 1. \tag{6.98}$$

By applying the three spheres inequality (6.51) over the balls centered at w_j, with radii ρ_j, $3\rho_j$ $4\rho_j$, for $j = 1, ..., k(t) - 1$, we are led to

$$\int_{B_{\rho_{k(t)}}(w_{k(t)})} |w|^2 \leq C\rho_0^3 \|\varphi\|^2_{H^{-\frac{1}{2}}(\partial\Omega,\mathbb{R}^n)} \tilde{\epsilon}^{2\beta_1 \delta^{k(t)-1}}, \tag{6.99}$$

where C only depends on α_0, β_0, M, α, M_0, M_1 and $\frac{\tilde{\rho}_0}{\rho_0}$.

From the interpolation inequality (6.64) and from (6.3) we have

$$\|w\|_{L^\infty(B_{\rho_{k(t)}}(w_{k(t)}),\mathbb{M}^n)} \leq \frac{C}{\rho_0^{\frac{n-3}{2}}} \|\varphi\|_{H^{-\frac{1}{2}}(\partial\Omega,\mathbb{R}^n)} \frac{\tilde{\epsilon}^{\beta_2 \delta^{k(t)-1}}}{\chi^{\frac{n}{2}(k(t)-1)}}, \tag{6.100}$$

where $\beta_2 = \frac{2\beta_1}{n+2}$ only depends on α_0, β_0, M, α, M_0, M_1 and $\frac{\tilde{\rho}_0}{\rho_0}$. Let us consider the point $z_t = z + t\xi$. We have that $z_t \in B_{\rho_{k(t)}}(w_{k(t)})$. From (6.100) and (6.3), we have that for any t, $0 < t \leq d(1)$,

$$|w(z)| \leq \frac{C}{\rho_0^{\frac{n-3}{2}}} \|\varphi\|_{H^{-\frac{1}{2}}(\partial\Omega,\mathbb{R}^n)} \left(\frac{t}{\rho_0} + \frac{\tilde{\epsilon}^{\beta_2 \delta^{k(t)-1}}}{\chi^{\frac{n}{2}(k(t)-1)}} \right). \tag{6.101}$$

Let

$$t(\tilde{\epsilon}) = d(1) \left|\log \tilde{\epsilon}^{\beta_2}\right|^{-\gamma},$$

with

$$\gamma = \frac{|\log \chi|}{2|\log \delta|}.$$

Let $\tilde{\mu} = \exp(-\beta_2^{-1})$. We have that $t(\tilde{\mu}) = d(1)$ and $t(\tilde{\epsilon}) \leq d(1)$ for any $\tilde{\epsilon}$, $0 < \tilde{\epsilon} \leq \tilde{\mu}$. Choosing $t = t(\tilde{\epsilon})$ in (6.101) and recalling (6.94) and (6.98), we have

$$\int_{\Omega_1 \setminus G} |\widehat{\nabla} u_1|^2 \leq C\rho_0 \|\varphi\|^2_{H^{-\frac{1}{2}}(\partial\Omega,\mathbb{R}^n)} \left|\log \tilde{\epsilon}^{\beta_2}\right|^{-B}, \tag{6.102}$$

where C only depends on α_0, β_0, M, α, M_0 and $\frac{\tilde{\rho}_0}{\rho_0}$. Therefore (4.2) and (4.5) follow.

In order to treat the case $\overline{r} \neq 0$, let us notice that

$$\partial(\Omega_1 \setminus G) = \Gamma_1 \cup \Gamma_2,$$

where $\Gamma_1 \subset \partial D_1$ and $\Gamma_2 = \partial D_2 \cap \partial G$.

When $n = 2$ we can distinguish the following three cases:

I) $\partial D_1 \cap \Gamma_2 = \emptyset$;

II) there exist at least two points z_1 and z_2, $z_i \in \partial D_1 \cap \Gamma_2$, $i = 1, 2$, satisfying

$$|z_1 - z_2| \geq \rho_0 \left|\log \tilde{\epsilon}\right|^{-\frac{\gamma}{2}}; \tag{6.103}$$

III) $\operatorname{diam}(\partial D_1 \cap \Gamma_2) \leq \rho_0 \left|\log \tilde{\epsilon}\right|^{-\frac{\gamma}{2}}$,

and then argue similarly to the proof of Proposition 4.2, up to the obvious changes.

The three dimensional case can be treated analogously, following the arguments used in the proof of Proposition 4.2 in Chapter 7. □

REMARK 6.6. Let us notice that Proposition 4.2 and Proposition 4.3 hold true also when condition (2.13) is removed and the traction field φ has support contained in the exterior boundary $\partial\Omega^e$ of Ω. In fact the only change in the proof of these Propositions is the appearance of new addends in the form of integrals over portions of the interior boundary $\partial\Omega^i = \partial\Omega \backslash \partial\Omega^e$. The integrand function of these integrals is a scalar product in which the conormal derivative of u_i is one of the two factors, see for instance (6.77) and therefore they vanish since we have assumed homogeneous Neumann condition on $\partial\Omega^i$.

CHAPTER 7

Proof of Proposition 4.2 in the 3-D case

In this last Chapter we shall prove Proposition 4.2 when the infinitesimal rigid displacement $\overline{r}$ appearing in (2.38) is different from zero and $n = 3$. Since the proof is mainly based on arguments which are analogous to those used in the two dimensional case, we shall only indicate the changes needed for the treatment of the three dimensional case.

PROOF OF PROPOSITION 4.2 FOR $\overline{r} \neq 0$ AND $n = 3$. It is convenient, similarly to the two–dimensional case, to distinguish the following three cases:

I) $\partial \tilde{D}_1^{\tilde{\rho}} \cap \tilde{\Gamma}_2^{\tilde{\rho}} = \emptyset$;

II) there exist at least three points z_1, z_2 and z_3, $z_i \in \partial \tilde{D}_1^{\tilde{\rho}} \cap \tilde{\Gamma}_2^{\tilde{\rho}}$, $i = 1, 2, 3$, such that the triangle $\Delta(z_1, z_2, z_3)$ having vertices z_1, z_2, z_3 satisfies the following inequality

$$\text{area}(\Delta(z_1, z_2, z_3)) \geq \rho_0^2 \left(\log|\log \tilde{\epsilon}|\right)^{-\frac{1}{4n}}; \tag{7.1}$$

III) for every triangle Δ having vertices belonging to $\partial \tilde{D}_1^{\tilde{\rho}} \cap \tilde{\Gamma}_2^{\tilde{\rho}}$ one has

$$\text{area}(\Delta) \leq \rho_0^2 \left(\log|\log \tilde{\epsilon}|\right)^{-\frac{1}{4n}}. \tag{7.2}$$

When I) holds the same proof given for $n = 2$ works in this case. In case II), the estimate (6.87) of $\overline{r}$ has to be modified as follows. By using the same arguments seen in the two dimensional case, we obtain

$$|\overline{r}(z_j)| \leq C \|\varphi\|_{H^{-\frac{1}{2}}(\partial\Omega, \mathbb{R}^n)} \left(\log|\log \tilde{\epsilon}|\right)^{-\frac{1}{n}}, \quad \text{j=1,2,3}, \tag{7.3}$$

where $C > 0$ only depends on α_0, β_0, M, α, M_0 and M_1.

Setting $\overline{r}(x) = c + a \times x$, where c, $a \in \mathbb{R}^3$, by (7.1) and (7.3) we have

$$|a||z_1 - z_2||\sin \varphi(a, z_1 - z_2)| = |\overline{r}(z_1) - \overline{r}(z_2)| \leq C \|\varphi\|_{H^{-\frac{1}{2}}(\partial\Omega, \mathbb{R}^n)} \left(\log|\log \tilde{\epsilon}|\right)^{-\frac{1}{n}}, \tag{7.4}$$

$$|a||z_1 - z_3||\sin \varphi(a, z_1 - z_3)| = |\overline{r}(z_1) - \overline{r}(z_3)| \leq C \|\varphi\|_{H^{-\frac{1}{2}}(\partial\Omega, \mathbb{R}^n)} \left(\log|\log \tilde{\epsilon}|\right)^{-\frac{1}{n}}, \tag{7.5}$$

$$|z_1 - z_2||z_1 - z_3||\sin \varphi(z_1 - z_2, z_1 - z_3)| = 2\,\text{area}(\Delta(z_1, z_2, z_3)) \geq 2\rho_0^2 \left(\log|\log \tilde{\epsilon}|\right)^{-\frac{1}{4n}}, \tag{7.6}$$

where $C > 0$ only depends on α_0, β_0, M, α, M_0 and M_1, and $\varphi(v, w)$ denotes the angle between the vectors v and w.

Since

$$|z_i - z_j| \leq C\rho_0, \quad i, j = 1, 2, 3, \tag{7.7}$$

by (7.6) we have

$$|z_1 - z_2| \geq C\rho_0 \left(\log|\log\tilde{\epsilon}|\right)^{-\frac{1}{4n}}, \tag{7.8}$$

$$|z_1 - z_3| \geq C\rho_0 \left(\log|\log\tilde{\epsilon}|\right)^{-\frac{1}{4n}}, \tag{7.9}$$

where $C > 0$ only depends on α_0, β_0, M, α, M_0 and M_1.

Let us begin by estimating $|a|$. By inserting (7.8) in (7.4) and (7.9) in (7.5), respectively, we have

$$|a||\sin\varphi(a, z_1 - z_2)| \leq C\frac{\|\varphi\|_{H^{-\frac{1}{2}}(\partial\Omega,\mathbb{R}^n)}}{\rho_0} \left(\log|\log\tilde{\epsilon}|\right)^{-\frac{3}{4n}}, \tag{7.10}$$

$$|a||\sin\varphi(a, z_1 - z_3)| \leq C\frac{\|\varphi\|_{H^{-\frac{1}{2}}(\partial\Omega,\mathbb{R}^n)}}{\rho_0} \left(\log|\log\tilde{\epsilon}|\right)^{-\frac{3}{4n}}, \tag{7.11}$$

where $C > 0$ only depends on α_0, β_0, M, α, M_0 and M_1.

If either $|\sin\varphi(a, z_1 - z_2)| \geq \left(\log|\log\tilde{\epsilon}|\right)^{-\frac{3}{8n}}$ or $|\sin\varphi(a, z_1 - z_3)| \geq \left(\log|\log\tilde{\epsilon}|\right)^{-\frac{3}{8n}}$, then either by (7.10) or by (7.11) we have

$$|a| \leq C\frac{\|\varphi\|_{H^{-\frac{1}{2}}(\partial\Omega,\mathbb{R}^n)}}{\rho_0} \left(\log|\log\tilde{\epsilon}|\right)^{-\frac{3}{8n}}, \tag{7.12}$$

where $C > 0$ only depends on α_0, β_0, M, α, M_0 and M_1.

It remains, therefore, to consider the case

$$\begin{cases} |\sin\varphi(a, z_1 - z_2)| \leq \left(\log|\log\tilde{\epsilon}|\right)^{-\frac{3}{8n}}, \\ |\sin\varphi(a, z_1 - z_3)| \leq \left(\log|\log\tilde{\epsilon}|\right)^{-\frac{3}{8n}}. \end{cases} \tag{7.13}$$

By (7.13) it is obvious that

$$|\sin\varphi(z_1 - z_2, z_1 - z_3)| \leq C\left(\log|\log\tilde{\epsilon}|\right)^{-\frac{3}{8n}}, \tag{7.14}$$

where $C > 0$ only depends on α_0, β_0, M, α, M_0 and M_1. By (7.6), (7.7) and (7.14) we have

$$1 \leq C\left(\log|\log\tilde{\epsilon}|\right)^{-\frac{1}{8n}}, \tag{7.15}$$

where $C > 0$ only depends on α_0, β_0, M, α, M_0 and M_1, which gives a contradiction for $\tilde{\epsilon}$ small enough.

By (7.3) and (7.12) we have

$$|c| \leq C\|\varphi\|_{H^{-\frac{1}{2}}(\partial\Omega,\mathbb{R}^n)} \left(\log|\log\tilde{\epsilon}|\right)^{-\frac{3}{8n}}, \tag{7.16}$$

where $C > 0$ only depends on α_0, β_0, M, α, M_0 and M_1, and finally

$$\|\overline{r}\|_{L^\infty(\Omega)} \leq C\|\varphi\|_{H^{-\frac{1}{2}}(\partial\Omega,\mathbb{R}^n)} \left(\log|\log\tilde{\epsilon}|\right)^{-\frac{3}{8n}}, \tag{7.17}$$

where $C > 0$ only depends on α_0, β_0, M, α, M_0 and M_1.

When III) holds, we only need to modify the proof of the estimate of the third addend in the right hand side of (6.89).

If there exist two points $w_1, w_2 \in \partial\tilde{D}_1^{\tilde{\rho}} \cap \tilde{\Gamma}_2^{\tilde{\rho}}$ such that

$$|w_1 - w_2| > \rho_0 \left(\log|\log\tilde{\epsilon}|\right)^{-\frac{1}{8n}}, \tag{7.18}$$

then, by (7.2) $\partial\tilde{D}_1^{\tilde{\rho}}\cap\tilde{\Gamma}_2^{\tilde{\rho}}$ is contained in the intersection between $\partial\tilde{D}_1^{\tilde{\rho}}$ and the open cylinder $\mathcal{C}$ having as axis the line l connecting w_1 and w_2, and radius $\eta = 2\rho_0\left(\log|\log\tilde{\epsilon}|\right)^{-\frac{1}{8n}}$.

Otherwise, $\mathrm{diam}(\partial\tilde{D}_1^{\tilde{\rho}}\cap\tilde{\Gamma}_2^{\tilde{\rho}}) \leq \rho_0\left(\log|\log\tilde{\epsilon}|\right)^{-\frac{1}{8n}}$, and it is easy to see that in this case $\partial\tilde{D}_1^{\tilde{\rho}}\cap\tilde{\Gamma}_2^{\tilde{\rho}}\subset B_{\frac{\eta}{2}}(z)$, for some $z\in\mathbb{R}^3$. By choosing arbitrarily a line l passing through z, we have that $\partial\tilde{D}_1^{\tilde{\rho}}\cap\tilde{\Gamma}_2^{\tilde{\rho}}$ is contained in the open cylinder having axis l and radius η. Therefore in both cases there exists an open cylinder $\mathcal{C}$ having axis l and radius η such that $\partial\tilde{D}_1^{\tilde{\rho}}\cap\tilde{\Gamma}_2^{\tilde{\rho}}\subset\partial\tilde{D}_1^{\tilde{\rho}}\cap\mathcal{C}$.

At this stage it would be desiderable to have the analogous of Lemma 3.1 in the present context, that is, that $\partial\tilde{D}_1^{\tilde{\rho}}\setminus\mathcal{C}$ is connected. However, this cannot hold in general, as it is easy to show by constructing simple counterexamples. What we can indeed prove is the weaker result stated in the Lemma 7.1 below, which will be proved at the end of this Chapter.

Let us recall that, by Lemma 6.3, $\partial\tilde{D}_1^{\tilde{\rho}}$ has boundary of class C^1 with constants $\tilde{\rho}_0$ and $\tilde{M}_0$, with the ratios $\frac{\tilde{\rho}_0}{\rho_0}$ and $\frac{\tilde{M}_0}{M_0}$ only depending on M_0 and α. Let us denote $\tilde{\beta} = \arctan\tilde{M}_0$ and notice that $\cos\tilde{\beta} = (1+\tilde{M}_0^2)^{-\frac{1}{2}}$. From now on let us choose ϵ small enough to ensure that $2\eta\sqrt{1+\tilde{M}_0^2}\leq\frac{\tilde{\rho}_0\cos^2\tilde{\beta}}{2}$, that is, equivalently, $\eta\leq\frac{\tilde{\rho}_0}{4(1+\tilde{M}_0^2)^{\frac{3}{2}}}$.

Let us denote by $\mathcal{C}^*$ the open cylinder having axis l and radius $\eta\sqrt{1+\tilde{M}_0^2}$.

LEMMA 7.1. *Let $\tilde{D}$ be a domain in $\mathbb{R}^3$ having boundary $\partial\tilde{D}$ connected, of Lipschitz class with constants $\tilde{\rho}_0$, $\tilde{M}_0$, and satisfying* $\mathrm{area}(\partial\tilde{D})\leq M_2\tilde{\rho}_0^2$. *Let $\mathcal{C}$ and $\mathcal{C}^*$ be open cylinders having a common axis l and radius η and $\eta\sqrt{1+\tilde{M}_0^2}$, respectively, with $\eta\leq\frac{\tilde{\rho}_0}{4(1+\tilde{M}_0^2)^{\frac{3}{2}}}$. Given any two points P_0, $Q_0\in\partial\tilde{D}\setminus\mathcal{C}^*$, there exists a path connecting P_0 and Q_0 inside $\partial\tilde{D}\setminus\mathcal{C}$. Moreover,*

$$\mathrm{area}(\partial\tilde{D}\cap\mathcal{C}^*)\leq C\tilde{\rho}_0\eta, \tag{7.19}$$

where $C>0$ only depends on $\tilde{M}_0$ and M_2.

Let us notice that, by Lemma 6.3, the hypotheses of the above Lemma hold for $\tilde{D}_1^{\tilde{\rho}}$, with M_2 only depending on M_0, M_1 and α. We have that any path on $\partial\tilde{D}_1^{\tilde{\rho}}$ connecting a point of $\tilde{\Gamma}_1^{\tilde{\rho}}$ with a point of $\partial\tilde{D}_1^{\tilde{\rho}}\setminus\tilde{\Gamma}_1^{\tilde{\rho}}$ must intersect $\partial\tilde{D}_1^{\tilde{\rho}}\cap\tilde{\Gamma}_2^{\tilde{\rho}}\subset\partial\tilde{D}_1^{\tilde{\rho}}\cap\mathcal{C}$. If $\partial\tilde{D}_1^{\tilde{\rho}}\setminus\mathcal{C}^*$ contains a point $P\in\tilde{\Gamma}_1^{\tilde{\rho}}$ and a point $Q\in\partial\tilde{D}_1^{\tilde{\rho}}\setminus\tilde{\Gamma}_1^{\tilde{\rho}}$, we know, by the above Lemma, that there exists a path γ joining P and Q inside $\partial\tilde{D}_1^{\tilde{\rho}}\setminus\mathcal{C}$, so that γ does not intersect $\partial\tilde{D}_1^{\tilde{\rho}}\cap\tilde{\Gamma}_2^{\tilde{\rho}}$, leading to a contradiction. Therefore we have that either $\tilde{\Gamma}_1^{\tilde{\rho}}\subset\partial\tilde{D}_1^{\tilde{\rho}}\cap\mathcal{C}^*$ or $\tilde{\Gamma}_1^{\tilde{\rho}}\supset\partial\tilde{D}_1^{\tilde{\rho}}\setminus\mathcal{C}^*$.

If $\tilde{\Gamma}_1^{\tilde{\rho}}\subset\partial\tilde{D}_1^{\tilde{\rho}}\cap\mathcal{C}^*$, then the third integral in the right hand side of (6.89) is easily estimated by recalling (6.2), (7.19) and the choice of $\eta = 2\rho_0(\log|\log\tilde{\epsilon}|)^{-\frac{1}{8n}}$. If, otherwise, $\tilde{\Gamma}_1^{\tilde{\rho}}\supset\partial\tilde{D}_1^{\tilde{\rho}}\setminus\mathcal{C}^*$, by (6.91) we have

$$-\int_{\tilde{\Gamma}_1^{\tilde{\rho}}}(\mathbb{C}\nabla u_1)\nu\cdot\overline{r} = \int_{\partial\tilde{D}_1^{\tilde{\rho}}\setminus\tilde{\Gamma}_1^{\tilde{\rho}}}(\mathbb{C}\nabla u_1)\nu\cdot\overline{r}\leq\int_{\partial\tilde{D}_1^{\tilde{\rho}}\cap\mathcal{C}^*}|(\mathbb{C}\nabla u_1)\nu\cdot\overline{r}|, \tag{7.20}$$

so that we reduce to the previous case. □

PROOF OF LEMMA 7.1. Let us fix an orientation on the axis l of the cylinder $\mathcal{C}^*$ which we shall refer to as the positive orientation of l. Let $\Pi_l : \mathbb{R}^3 \to l$ be the orthogonal projection on the line l and let us introduce in $\mathbb{R}^3$ the ordering induced by the positive orientation of l as follows

$$P \text{ follows } Q \Longleftrightarrow \Pi_l(P) \text{ follows } \Pi_l(Q) \text{ on } l. \tag{7.21}$$

We have that $\Pi_l(\partial\tilde{D} \cap \overline{\mathcal{C}}^*)$ is a compact subset of l, whose connected closure is a closed segment S having a starting point P^S and an ending point P^E both belonging to $\Pi_l(\partial\tilde{D} \cap \overline{\mathcal{C}}^*)$.

Given any two points P_0, $Q_0 \in \partial\tilde{D} \setminus \mathcal{C}^*$, we know, by the arguments used in the proof of Lemma 3.1, that there exists a path $\gamma : [a, b] \to \partial\tilde{D}$ joining P_0 and Q_0 such that length$(\gamma) \leq K\tilde{\rho}_0$, with $K > 0$ only depending on $\tilde{M}_0$ and M_2.

Now, if $\gamma([a, b]) \subset \partial\tilde{D} \setminus \mathcal{C}$, then we are done. Otherwise, let us show how to modify γ to obtain the thesis.

Let us consider the closed, nonempty set

$$J = \{t \in (a, b) \mid \gamma(t) \in \partial\tilde{D} \cap \mathcal{C}^*\}. \tag{7.22}$$

and let us define

$$t_{\min} = \min J, \quad R_{\min} = \gamma(t_{\min}). \tag{7.23}$$

Let us notice that, by the continuity of the map γ, $R_{\min} \in \partial\tilde{D} \cap \partial\mathcal{C}^*$.

Claim. There exist $\hat{t} \in (t_{\min}, b]$ such that $\gamma(\hat{t}) \in \partial\tilde{D} \setminus \mathcal{C}^*$ and a path $\hat{\gamma}$ connecting $R_{\min} = \gamma(t_{\min})$ with $\gamma(\hat{t})$ inside $\partial\tilde{D} \setminus \mathcal{C}$ such that either $\hat{t} = b$ or length$(\gamma|_{[t_{\min}, \hat{t}]}) \geq \tilde{\rho}_0 \cos\tilde{\beta}$.

PROOF OF THE CLAIM. There exists a rigid transformation of coordinates under which we have $R_{\min} = 0$ and

$$\partial\tilde{D} \cap B_{\tilde{\rho}_0}(0) = \{x = (x', x_3) \in B_{\tilde{\rho}_0}(0) \mid x_3 = \psi(x')\}, \tag{7.24}$$

where ψ is a Lipschitz function defined on the disk $B'_{\tilde{\rho}_0}(0)$ in the plane $Ox'_1x'_2$ satisfying

$$\psi(0) = 0, \quad \|\psi\|_{C^{0,1}(B'_{\tilde{\rho}_0}(0))} \leq \tilde{M}_0\tilde{\rho}_0. \tag{7.25}$$

Let us notice that the restriction of the graph of ψ to the disk $\overline{B'_{\tilde{\rho}_0 \cos\tilde{\beta}}(0)}$ is contained in $\partial\tilde{D}$.

Let us denote by Π the projection on the plane $Ox'_1x'_2$. If $|Q_0 - \gamma(t_{\min})| > \tilde{\rho}_0 \cos\tilde{\beta}$, then let us define $t_1 = \max\{t \in (t_{\min}, b] \mid |\gamma(t) - \gamma(t_{\min})| = \tilde{\rho}_0 \cos\tilde{\beta}\}$, otherwise let $t_1 = b$. It is evident that if $t_1 < b$, that is if $\gamma(t_1) \neq Q_0$, the length of $\gamma|_{[t_{\min}, t_1]}$ is at least $\tilde{\rho}_0 \cos\tilde{\beta}$.

Let

$$Z = \{x' \in \overline{B'_{\tilde{\rho}_0 \cos\tilde{\beta}}} \mid (x', \psi(x')) \in \partial\tilde{D} \cap \mathcal{C}^*\}. \tag{7.26}$$

We have that $0 \notin Z$ and that Z is contained in $\Pi(\mathcal{C}^*)$.

Let us distinguish two cases:

I) The common axis l of the cylinders $\mathcal{C}$ and $\mathcal{C}^*$ is orthogonal to the plane $Ox'_1x'_2$. In this case $\Pi(\mathcal{C})$ and $\Pi(\mathcal{C}^*)$ are concentric disks of radii η and $\eta\sqrt{1 + \tilde{M}_0^2}$ respectively, such that the origin belongs to the boundary of $\Pi(\mathcal{C}^*)$.

II) The axis l is not orthogonal to the plane $Ox'_1x'_2$. In this case $\Pi(\mathcal{C})$ and $\Pi(\mathcal{C}^*)$ are parallel strips having the common middle line $\Pi(l)$, width 2η and $2\eta\sqrt{1+\tilde{M}_0^2}$ respectively, such that the origin belongs to $\overline{\Pi(\mathcal{C}^*)}$. Let us remark that the origin may belong to the thinner strip $\Pi(\mathcal{C})$ and even to the middle line $\Pi(l)$.

In case I), recalling that $2\eta\sqrt{1+\tilde{M}_0^2} \leq \frac{\tilde{\rho}_0 \cos^2\tilde{\beta}}{2} \leq \frac{\tilde{\rho}_0 \cos\tilde{\beta}}{2}$, it follows that the disk $\Pi(\mathcal{C}^*)$ is compactly contained in $B'_{\tilde{\rho}_0\cos\tilde{\beta}}$. Moreover, for any x' belonging to the disk $\Pi(\mathcal{C}^*)$, we have that

$$|\psi(x')| \leq M_0|x'|, \quad |(x', \psi(x'))| \leq \frac{\tilde{\rho}_0 \cos^2\tilde{\beta}}{2}\sqrt{1+M_0^2} = \frac{\tilde{\rho}_0\cos\tilde{\beta}}{2}.$$

It follows that if $|\gamma(t_1) - \gamma(t_{\min})| = \tilde{\rho}_0\cos\tilde{\beta}$, then $\Pi(\gamma(t_1))$ does not belong to the disk $\Pi(\mathcal{C}^*)$. If, instead, $|\gamma(t_1)-\gamma(t_{\min})| < \tilde{\rho}_0\cos\tilde{\beta}$, then $t_1 = b$ and $\gamma(t_1) = Q_0 \notin \mathcal{C}^*$, so that, obviously, again, $\Pi(\gamma(t_1))$ does not belong to the disk $\Pi(\mathcal{C}^*)$.

Therefore, in both cases, we can construct a path σ joining $\Pi(\gamma(t_{\min})) = 0$ with $\Pi(\gamma(t_1))$ inside $B'_{\tilde{\rho}_0\cos\tilde{\beta}} \setminus Z$. The path $(\sigma, \psi\circ\sigma)$ joins $\gamma(t_{\min})$ with $\gamma(t_1)$ inside $\partial\tilde{D} \setminus \mathcal{C}^*$.

In this case the thesis follows with $\hat{t} = t_1$.

In case II), since $2\eta\sqrt{1+\tilde{M}_0^2} \leq \frac{\tilde{\rho}_0 \cos^2\tilde{\beta}}{2} \leq \frac{\tilde{\rho}_0 \cos\tilde{\beta}}{2}$, it is clear that the set $\overline{B'_{\tilde{\rho}_0\cos\tilde{\beta}}} \setminus \Pi(\mathcal{C}^*)$ has exactly two connected components, which we denote by E_+, E_-, and that and $\partial(\Pi(\mathcal{C}^*)) \cap \overline{B'_{\tilde{\rho}_0\cos\tilde{\beta}}}$ consists of two closed segments, which we denote by S^*_+, S^*_-, contained in E_+, E_- respectively and which inherit the orientation of the line l as described above. The distance from the origin of at least one among the segments S^*_+, S^*_-, say for instance S^*_+, does not exceed $\eta\sqrt{1+\tilde{M}_0^2}$.

If $\Pi(\gamma(t_{\min}))$ and $\Pi(\gamma(t_1))$ belong to the same connected component of the set $\overline{B'_{\tilde{\rho}_0\cos\tilde{\beta}}} \setminus Z$, we can construct a path σ joining the two points inside $\overline{B'_{\tilde{\rho}_0\cos\tilde{\beta}}} \setminus Z$ and then the path $(\sigma, \psi\circ\sigma)$ joins $\gamma(t_{\min})$ and $\gamma(t_1)$ inside $\partial\tilde{D} \setminus \mathcal{C}^*$.

Otherwise, let us consider the plane π orthogonal to the axis l and containing $\gamma(t_{\min})$. The plane π intersects the plane $Ox'_1x'_2$ into the line $l^\perp$ orthogonal to $\Pi(l)$ and passing through $\Pi(\gamma(t_{\min}))$ which coincides with the origin. The intersection of the cylinder $\mathcal{C}$ with the plane π is a disk of radius η having the center at the distance $\eta\sqrt{1+\tilde{M}_0^2}$ from the origin.

By representing in the plane π the line $l^\perp$, the graph of ψ and the disk $\mathcal{C}\cap\pi$, it is immediate to verify that the graph of ψ does not intersect $\mathcal{C}\cap\pi$ if one moves along $l^\perp$ from the origin to the point $S^*_+\cap\pi$.

We can construct a rectilinear path σ^1_+ joining $\Pi(\gamma(t_{\min})) = 0$ with the point $S^*_+\cap l^\perp$ (this step being unnecessary when $\Pi(\gamma(t_{\min})) \in S^*_+$), a rectilinear path σ^2_+ joining $S^*_+\cap l^\perp$ with the endpoint P_1 of S^*_+ with respect to the positive orientation, and, by gluing, a path σ_+ joining $\Pi(\gamma(t_{\min}))$ with P_1, such that the path $(\sigma_+, \psi\circ\sigma_+)$ joins $\gamma(t_{\min})$ with $(P_1, \psi(P_1))$ inside $\partial\tilde{D} \setminus \mathcal{C}$.

Let us distinguish again two cases:

IIa) $\gamma(t_1) \in \partial\tilde{D} \setminus \mathcal{C}^*$;

IIb) $\gamma(t_1) \in \partial\tilde{D} \cap \mathcal{C}^*$.

Let us consider first case IIa) and let us notice that this condition is certainly satisfied when $t_1 = b$. Since we are assuming that $\Pi(\gamma(t_{\min}))$ and $\Pi(\gamma(t_1))$ belong to different components of $\overline{B'_{\tilde{\rho}_0 \cos\tilde{\beta}}} \setminus Z$, we have that $\Pi(\gamma(t_1))$ belongs either to E_- or to $\Pi(\mathcal{C}^*)$. In both cases, arguing as above, we can construct a path σ_- joining $\Pi(\gamma(t_1))$ with the endpoint Q_1 of S^*_- with respect to the positive orientation and such that the path $(\sigma_-, \psi\circ\sigma_-)$ joins $\gamma(t_1)$ with $(Q_1, \psi(Q_1))$ inside $\partial\tilde{D}\setminus\mathcal{C}$.

The arc of the circle $\partial B'_{\tilde{\rho}_0\cos\tilde{\beta}}$ connecting Q_1 and P_1 inside $\overline{\Pi(\mathcal{C}^*)}$ contains at least a point $R_1 \in Z$, since, otherwise, $\Pi(\gamma(t_{\min}))$ and $\Pi(\gamma(t_1))$ could be connected inside $\overline{B'_{\tilde{\rho}_0\cos\tilde{\beta}}}\setminus Z$, leading to a contradiction.

Recalling that $2\eta\sqrt{1+\tilde{M}_0^2} \leq \frac{\tilde{\rho}_0\cos^2\tilde{\beta}}{2}$, by standard computation, we have that

$$\operatorname{dist}(R_1, l^\perp) \geq \frac{\sqrt{3}}{2}\tilde{\rho}_0\cos\tilde{\beta},$$

$$|R_1 - P_1| \leq \frac{\sqrt{2}}{2}\tilde{\rho}_0\cos^2\tilde{\beta},$$

$$|R_1 - Q_1| \leq \frac{\sqrt{2}}{2}\tilde{\rho}_0\cos^2\tilde{\beta},$$

and, consequently,

$$(R_1, \psi(R_1)) \in \partial\tilde{D}\cap\mathcal{C}^*,$$

$$|\Pi_l(R_1, \psi(R_1)) - \Pi_l(R_{\min})| \geq \frac{\sqrt{2}}{2}\tilde{\rho}_0\cos\tilde{\beta},$$

$$|(R_1, \psi(R_1)) - (P_1, \psi(P_1))| \leq \frac{\sqrt{2}}{2}\tilde{\rho}_0\cos\tilde{\beta},$$

$$|(R_1, \psi(R_1)) - (Q_1, \psi(Q_1))| \leq \frac{\sqrt{2}}{2}\tilde{\rho}_0\cos\tilde{\beta}.$$

Moreover we have that $\Pi_l(R_1, \psi(R_1))$ follows $\Pi_l(R_{\min})$ on S with respect to the positive orientation.

Given a local representation of $\partial\tilde{D}$ around $(R_1, \psi(R_1))$ of type (7.24), and still denoting by Π the projection on the plane $Ox'_1x'_2$ relative to this local representation, we trivially have that

$$\begin{aligned} &|\Pi(R_1, \psi(R_1)) - \Pi(P_1, \psi(P_1))| < \tilde{\rho}_0\cos\tilde{\beta}, \\ &|\Pi(R_1, \psi(R_1)) - \Pi(Q_1, \psi(Q_1))| < \tilde{\rho}_0\cos\tilde{\beta}. \end{aligned} \tag{7.27}$$

Now, if $\Pi(P_1, \psi(P_1))$ and $\Pi(Q_1, \psi(Q_1))$ belong to the same connected component of $\overline{B'_{\tilde{\rho}_0\cos\tilde{\beta}}}\setminus Z$, with Z given by (7.26) in the present local representation, then, by following previous arguments, we can join $(P_1, \psi(P_1)$ with $(Q_1, \psi(Q_1)$ inside $\partial\tilde{D}\setminus\mathcal{C}^*$ and therefore, by gluing of paths, we can connect $\gamma(t_{\min})$ with $\gamma(t_1)$ inside $\partial\tilde{D}\setminus\mathcal{C}$.

Otherwise, if $\Pi(P_1, \psi(P_1))$ and $\Pi(Q_1, \psi(Q_1))$ do not belong to the same connected component of $\overline{B'_{\tilde{\rho}_0\cos\tilde{\beta}}}\setminus Z$, then, by the arguments seen above in treating case I), it is clear that also in this new local representation case II) holds. Therefore we can repeat the above construction defining similarly points R_2, P_2, Q_2 and so on. By our regularity assumptions, length$(S) \leq K\tilde{\rho}_0$, with K only depending on $\tilde{M}_0$ and M_2. Moreover, at each step, the point $\Pi_l(R_j, \psi(R_j))$ follows $\Pi_l(R_{j-1}, \psi(R_{j-1}))$ on S with respect to the positive orientation, with

$$|\Pi_l(R_j, \psi(R_j)) - \Pi_l(R_{j-1}, \psi(R_{j-1}))| \geq \frac{\sqrt{2}}{2}\tilde{\rho}_0\cos\tilde{\beta}.$$

Therefore, in a finite number of steps we reduce to the case in which $\Pi(P_k, \psi(P_k))$ and $\Pi(Q_k, \psi(Q_k))$ belong to the same connected component of the set $\overline{B'_{\tilde\rho_0 \cos\tilde\beta}} \setminus Z$, for some k, so that, by gluing of paths, we can connect $\gamma(t_{\min})$ with $\gamma(t_1) \notin \mathcal{C}^*$ inside $\partial\tilde D \setminus \mathcal{C}$.

Also in this case the thesis follows with $\hat t = t_1$.

Let us consider now the case IIb), when $\gamma(t_1) \in \mathcal{C}^*$. In this case, necessarily, we have $|\gamma(t_1) - \gamma(t_{\min})| = \tilde\rho_0 \cos\tilde\beta$. By denoting $d_1 = |\Pi(\gamma(t_1)) - \Pi(\gamma(t_{\min}))| = |\Pi(\gamma(t_1))|$, we have that $\tilde\rho_0 \cos^2\tilde\beta \le d_1 \le \tilde\rho_0\cos\tilde\beta$. The set $\overline{\Pi(\mathcal{C}^*)} \cap \partial B'_{d_1}$ consists of two closed arcs, one of which, which we denote by Γ, contains $\Pi(\gamma(t_1))$. Let $P_1 = \Gamma \cap S^*_+$. Arguing similarly to above, we can construct a path σ_+ joining $\Pi(\gamma(t_{\min})) = 0$ with P_1 such that the path $(\sigma_+, \psi \circ \sigma_+)$ joins $\gamma(t_{\min})$ with $(P_1, \psi(P_1))$ inside $\partial\tilde D \setminus \mathcal{C}$. We can compute

$$|P_1 - \Pi(\gamma(t_1))| \le \frac{\sqrt{2}}{2}\tilde\rho_0 \cos^2\tilde\beta,$$

so that

$$|(P_1, \psi(P_1)) - \gamma(t_1)| \le \frac{\sqrt{2}}{2}\tilde\rho_0 \cos\tilde\beta.$$

Now, if $|Q_0 - \gamma(t_1)| > \tilde\rho_0\cos\tilde\beta$ then let us define

$$t_2 = \max\{t \in (t_1, b] \mid |\gamma(t) - \gamma(t_1)| = \tilde\rho_0\cos\tilde\beta\},$$

otherwise let us define $t_2 = b$.

Given a local representation of $\partial\tilde D$ around $\gamma(t_1)$ of type (7.24), and still denoting by Π the projection on the plane $Ox'_1x'_2$ relative to this local representation, we have that

$$|\Pi(P_1, \psi(P_1)) - \Pi(\gamma(t_1))| < \tilde\rho_0\cos\tilde\beta.$$

If $\Pi(P_1, \psi(P_1))$ and $\Pi(\gamma(t_2))$ belong to the same connected component of $\overline{B'_{\tilde\rho_0\cos\tilde\beta}} \setminus Z$ with Z given by (7.26) in the present local representation, then, by the previous arguments, we can join $(P_1, \psi(P_1)$ with $\gamma(t_2)$ inside $\partial\tilde D \setminus \mathcal{C}^*$ and, by gluing of paths, we can connect $\gamma(t_{\min})$ with $\gamma(t_2)$ inside $\partial\tilde D \setminus \mathcal{C}$.

Otherwise, we are either in case IIa) or in case IIb). If IIa) holds, then we just know how to connect $(P_1, \psi(P_1))$ with $\gamma(t_2)$ inside $\partial\tilde D \setminus \mathcal{C}$ and therefore we have a path connecting $\gamma(t_{\min})$ with $\gamma(t_2)$ inside $\partial\tilde D \setminus \mathcal{C}$. Moreover, $\text{length}(\gamma|_{[t_{\min}, t_2]}) \ge \text{length}(\gamma|_{[t_{\min}, t_1]}) \ge \tilde\rho_0\cos\tilde\beta$.

If, instead, IIb) holds, we can repeat the above construction defining similarly a point P_2 such that $(P_1, \psi(P_1))$ and $(P_2, \psi(P_2))$ can be connected inside $\partial\tilde D \setminus \mathcal{C}^*$, and so on. We can repeat iteratively the above construction starting from $\gamma(t_2)$ and P_2. Since $\text{length}(\gamma|_{[t_i, t_{i+1}]}) \ge \tilde\rho_0\cos\tilde\beta$ and since $\gamma(b) = Q_0 \notin \mathcal{C}^*$, in a finite number of steps we must have $\gamma(t_k) \notin \mathcal{C}^*$, so that, by the analysis of cases I) and IIa), we can connect $\gamma(t_k)$ with $(P_{k-1}, \psi(P_{k-1}))$ and therefore with $\gamma(t_{\min})$ inside $\partial\tilde D \setminus \mathcal{C}$.

In this case the thesis follows with $\hat t = t_k$. □

By replacing the given path γ in the interval $[t_{\min}, \hat t]$ with the path $\hat\gamma$ obtained by the Claim, and still denoting the path so modified by γ, we have that either $\gamma([\hat t, b]) \subset \partial\tilde D \setminus \mathcal{C}$ or not. In the former case we are done, in the latter one we can repeat the construction of the Claim, and so on. Since $\text{length}(\gamma) \le K\tilde\rho_0$, with

$K > 0$ only depending on $\tilde{M}_0$ and M_2, in a finite number of steps we obtain the thesis.

It remains to prove (7.19).

Let $h = \frac{\sqrt{3}}{2}\tilde{\rho}_0$. Let us subdivide the segment S into N closed nonoverlapping segments S_i, having endpoints Z_{i-1}, Z_i, for $i = 1, ..., N$, where $Z_0 = P^S$ and $|Z_i - Z_{i-1}| = h$, for $i = 1, ..., N-1$, $0 < |Z_N - Z_{N-1}| \leq h$, $Z_N = P^E$. Recalling that length$(S) \leq K\tilde{\rho}_0$, with K only depending on $\tilde{M}_0$ and M_2, we have that $N \leq \frac{2K}{\sqrt{3}} + 1$. For every $i = 1, ..., N$, let us consider the truncated cylinder

$$\overline{C_i^*} = \{P \in \overline{C^*} \mid |\Pi_l(P) \in S_i\}.$$

By our choice of h and recalling that $\eta \leq \frac{\tilde{\rho}_0}{4(1+\tilde{M}_0^2)^{\frac{3}{2}}}$, we have that diam$(\overline{C_i^*}) \leq \tilde{\rho}_0$. Moreover $\partial\tilde{D} \cap C^* \subset \cup_{i=1}^N \partial\tilde{D} \cap \overline{C_i^*}$.

Let us assume that $Q \in \partial\tilde{D} \cap \overline{C_j^*}$ for some $j \in \{1, ..., N\}$.

There exists a rigid transformation of coordinates under which $Q = 0$ and (7.24)–(7.25) hold.

Denoting, as usual, the projection on the plane $Ox_1'x_2'$ by Π, we have that

$$\partial\tilde{D} \cap \overline{C_j^*} \subset \{x = (x', x_3) \in B_{\tilde{\rho}_0}(0) \mid x' \in \Pi(\overline{C_j^*}),\ x_3 = \psi(x')\},$$

Now, $\Pi(\overline{C_j^*})$ is either a disk of radius $\eta\sqrt{1 + \tilde{M}_0^2}$ or is contained in the intersection of the disk $B'_{\tilde{\rho}_0}(0)$ with a strip of width $2\eta\sqrt{1 + \tilde{M}_0^2}$. Therefore, in both cases, area$(\partial\tilde{D} \cap \overline{C_j^*}) \leq (1 + \tilde{M}_0^2)^{\frac{1}{2}}$area$(\Pi(\overline{C_j^*})) \leq C\eta\tilde{\rho}_0$, with C only depending on $\tilde{M}_0$. Finally, recalling the bound for N, (7.19) follows. □

CHAPTER 8

A related inverse problem in electrostatics

In this Chapter we consider the related problem arising in the electrostatic context, consisting in determining, inside an electrical conductor Ω, an inclusion D made of perfectly conducting material, from a single measurement of current density and voltage potential taken at the boundary of Ω. If we apply a current flux φ at the boundary of Ω, then the induced potential inside Ω satisfies the following boundary value problem

$$\begin{cases} \operatorname{div}(\sigma\nabla u) = 0, & \text{in } \Omega\setminus\overline{D}, \\ \sigma\nabla u\cdot\nu = \varphi, & \text{on } \partial\Omega, \\ u_{|\partial D} \equiv const., \end{cases} \tag{8.1}$$

coupled with the *equilibrium condition*

$$\int_{\partial D} \sigma\nabla u\cdot\nu = 0, \tag{8.2}$$

where $\sigma = \{\sigma_{ij}(x)\}_{i,j=1}^n$ denotes the known symmetric conductivity tensor. Problem (8.1)–(8.2) admits a unique solution $u \in H^1(\Omega\setminus\overline{D})$ up to an additive constant. In order to specify a single solution, we shall assume, from now on, the following normalization condition

$$u = 0 \quad \text{on } \partial D. \tag{8.3}$$

We shall make the following a priori assumptions.

i) Assumptions about the boundary data.

On the Neumann data φ appearing in problem (8.1) we assume that

$$\varphi \in H^{-\frac{1}{2}}(\partial\Omega), \quad \varphi \not\equiv 0, \tag{8.4}$$

the (obvious) compatibility condition

$$\int_{\partial\Omega} \varphi = 0, \tag{8.5}$$

and that, for a given constant $F > 0$,

$$\frac{\|\varphi\|_{H^{-\frac{1}{2}}(\partial\Omega)}}{\|\varphi\|_{H^{-1}(\partial\Omega)}} \le F. \tag{8.6}$$

ii) Assumptions about the conductivity tensor.

The conductivity σ is assumed to be a given function defined in $\overline{\Omega}$ with values $n\times n$ symmetric matrices satisfying the following conditions for given constants λ, Λ, $0<\lambda\le 1$, $\Lambda \ge 0$,

$$\lambda|\xi|^2 \leq \sigma(x)\xi\cdot\xi \leq \lambda^{-1}|\xi|^2, \quad \text{for every } x\in\overline{\Omega}, \xi\in\mathbb{R}^n, \quad (\textit{ellipticity}) \tag{8.7}$$

$$|\sigma(x)-\sigma(y)| \leq \Lambda\frac{|x-y|}{\rho_0}, \quad \text{for every } x,y\in\overline{\Omega}. \quad (\textit{Lipschitz continuity}) \tag{8.8}$$

In the sequel we shall consider the following boundary value problem of mixed type

$$\begin{cases} \operatorname{div}(\sigma\nabla u)=0, & \text{in } \Omega\setminus\overline{D}, \\ \sigma\nabla u\cdot\nu=\varphi, & \text{on } \partial\Omega, \\ u=0, & \text{on } \partial D, \end{cases} \tag{8.9}$$

coupled with the *no-flux condition*

$$\int_{\partial D}\sigma\nabla u\cdot\nu=0. \tag{8.10}$$

THEOREM 8.1 (Stability). *Let Ω be a domain satisfying* (2.12), (2.13) *and* (2.18). *Let D_i, $i=1,2$, be two connected open subsets of Ω satisfying* (2.14)–(2.16) *and such that ∂D_i is of class $C^{1,1}$ with constants ρ_0, M_0. Moreover, let Σ be an open portion of $\partial\Omega$ satisfying* (2.17) *and of class $C^{1,\alpha}$ with constants ρ_0, M_0. Let $u_i\in H^1(\Omega\setminus\overline{D_i})$ be the solution to* (8.9)–(8.10), *when $D=D_i$, $i=1,2$, and let* (8.4)–(8.8) *be satisfied. If, given $\epsilon>0$, we have*

$$\|u_1-u_2-\overline{c}\|_{L^2(\Sigma)} = \min_{c\in\mathbb{R}}\|u_1-u_2-c\|_{L^2(\Sigma)} \leq \rho_0^{\frac{n-1}{2}}\epsilon, \tag{8.11}$$

then we have

$$d_{\mathcal{H}}(\partial D_1,\partial D_2) \leq \rho_0\omega\left(\frac{\epsilon}{\rho_0^{\frac{3-n}{2}}\|\varphi\|_{H^{-\frac{1}{2}}(\partial\Omega)}}\right) \tag{8.12}$$

and

$$d_{\mathcal{H}}(\overline{D_1},\overline{D_2}) \leq \rho_0\omega\left(\frac{\epsilon}{\rho_0^{\frac{3-n}{2}}\|\varphi\|_{H^{-\frac{1}{2}}(\partial\Omega)}}\right), \tag{8.13}$$

where ω is an increasing continuous function on $[0,\infty)$ which satisfies

$$\omega(t)\leq C|\log t|^{-\eta}, \quad \textit{for every } t,\ 0<t<1, \tag{8.14}$$

and C, η, $C>0$, $0<\eta\leq 1$, are constants only depending on the a priori data M_0, α, M_1, F, λ, Λ.

PROOF. Let us briefly give a sketch for a proof, which can be obtained by merging the techniques used in [**4**] and in the present paper.

As a first step, noticing that the role taken by the infinitesimal rigid displacements in the elasticity framework is played here by the constant functions, we can prove stability estimates of continuation from Cauchy data analogous to those stated here in Proposition 4.2 and Proposition 4.3, by adapting to the simpler scalar context the geometrical arguments introduced in the present paper.

As a second step, in the scalar case we may take advantage of the validity of a doubling inequality at the boundary to obtain the optimal logarithmic estimates (8.12)–(8.14) by following the lines developed in [**4**]. □

Bibliography

[1] V. Adolfsson and L. Escauriaza, $C^{1,\alpha}$ *domains and unique continuation at the boundary.* Comm. Pure Appl. Math. **L** (1997), 935–969.

[2] V. Adolfsson, L. Escauriaza and C. Kenig, *Convex domains and unique continuation at the boundary.* Rev. Mat. Iberoamericana **11** (1995), 513–525.

[3] G. Alessandrini, *Stable determination of conductivity by boundary measurements.* Appl. Anal. **27** (1988), 153–172.

[4] G. Alessandrini, E. Beretta, E. Rosset and S. Vessella, *Optimal stability for inverse elliptic boundary value problems with unknown boundaries.* Ann. Scuola Norm. Sup. Pisa Cl. Sci. (4) **XXIX** (2000), 755–806.

[5] G. Alessandrini and A. Morassi, *Strong unique continuation for the Lamé system of elasticity.* Comm. Partial Differential Equations **26** (2001), 1787–1810.

[6] G. Alessandrini, A. Morassi and E. Rosset, *Detecting an inclusion in an elastic body by boundary measurements.* SIAM J. Math. Anal. **33**(6) (2002), 1247–1268.

[7] G. Alessandrini, A. Morassi and E. Rosset, *Detecting cavities by electrostatic boundary measurements.* Inverse Problems **18** (2002), 1333–1353.

[8] G. Alessandrini and L. Rondi, *Optimal stability for the inverse problem of multiple cavities.* J. Differential Equations **176** (2001), 356–386.

[9] G. Alessandrini and E. Rosset, *The inverse conductivity problem with one measurement: bounds on the size of the unknown object.* SIAM J. Appl. Math. **58**(4) (1998), 1060–1071.

[10] D.D. Ang, D.D. Trong and M. Yamamoto, *Identification of cavities inside two-dimensional heterogeneous isotropic elastic bodies.* J. Elasticity **56** (1999), 199–212.

[11] J.A. Barceló, T. Barceló and A. Ruiz, *Stability in the inverse conductivity problem in the plane for less regular conductivities.* J. Differential Equations **173** (2001), 231–270.

[12] E. Beretta and S. Vessella, *Stable determination of boundaries from Cauchy data.* SIAM J. Math. Anal. **30** (1998), 220–232.

[13] L. Borcea, *Electrical impedance tomography.* Inverse Problems **18** (2002), R99–R136.

[14] A.L. Bukhgeim, J. Cheng and M. Yamamoto, *Conditional stability in an inverse problem of determining a non-smooth boundary.* J. Math. Anal. Appl. **242** (2000), 57–74.

[15] A.P. Calderón, *On an inverse boundary value problem*, Seminar on Numerical Analysis and its Applications to Continuum Physics, Sociedad Brasileira de Matemàtica, Rio de Janeiro, 1980, pp. 65–73.

[16] S. Campanato, *Sui problemi al contorno per sistemi di equazioni differenziali lineari del tipo dell'elasticità - (parte I).* Ann. Scuola Norm. Sup. Pisa Cl. Sci. **XIII** (1959), 223–258.

[17] S. Campanato. Equazioni ellittiche del secondo ordine e spazi $\mathcal{L}^{2,\lambda}$. *Ann. Mat. Pura Appl.* **69** (1965) 321–382.

[18] S. Campanato, *Sistemi ellittici in forma divergenza. Regolarità all'interno*, Quaderni Scuola Normale Superiore Pisa, Pisa, 1980.

[19] J. Cheng, Y.C. Hon and M. Yamamoto, *Conditional stability estimation for an inverse boundary problem with non-smooth boundary in* $\mathbb{R}^3$. Trans. Amer. Math. Soc. **353** (2001), 4123–4138.

[20] B.E.J. Dahlberg and C.E. Kenig, L^p *estimates for the three-dimensional systems of elastostatics on Lipschitz domains*, in *Analysis and Partial Differential Equations*, Lecture Notes in Pure Appl. Math. vol 122, Dekker, New York, 1990, pp. 621–634.

[21] B.E.J. Dahlberg, C.E. Kenig and G.C. Verchota, *Boundary value problems for the systems of elastostactics in Lipschitz domains.* Duke Math. J. **57** (1988), 795–818.

[22] M. Di Cristo and L. Rondi, *Examples of exponential instability for inverse inclusion and scattering problems.* Inverse Problems **19** (2003), 685–701.

[23] M. Di Cristo and L. Rondi, *Examples of exponential instability for elliptic inverse problems.* Preprint arXiv (2003) (downloadable from http://arxiv.org/archive/math).

[24] M. Eller, V. Isakov, G. Nakamura, and D. Tataru, *Uniqueness and stability in the Cauchy problem for Maxwell and elasticity systems*, In *Nonlinear Partial Differential equations*, vol. 16, College de France Seminar, Chapman and Hill/CRC, 2000.

[25] G. Eskin and J. Ralston, *On the inverse boundary value problem for linear isotropic elasticity,* Inverse Problems **18** (2002), 907–921.

[26] W. Gao, *Layer potentials and boundary value problems for elliptic systems in Lipschitz domains.* J. Funct. Anal. **95** (1991), 377–399.

[27] M. Giaquinta and G. Modica, *Non linear systems of the type of the stationary Navier-Stokes system.* J. Reine Angew. Math. **330** (1982), 173–214.

[28] M. E. Gurtin, *The linear theory of elasticity*, in *Handbuch der Physik* VIa/2, 1–295, Springer, Berlin–Heidenberg–New York, 1972.

[29] V. Isakov, *Inverse problems for partial differential equations*, Applied Mathematical Sciences, vol. 127, Springer, New York, 1998.

[30] R. Kohn and M. Vogelius, *Determining conductivity by boundary measurements.* Comm. Pure Appl. Math. **37** (1984), 289–298.

[31] R. Kohn and M. Vogelius, *Determining conductivity by boundary measurements II. Interior results.* Comm. Pure Appl. Math. **38** (1985), 643–667.

[32] V.A. Kondrat'ev and O.A. Oleinik, *On the dependence of the constant in Korn's inequality on parameters characterizing the geometry of the region.* Russian Math. Surveys **44** (1989), 187–195.

[33] I. Kukavica and K. Nyström, *Unique continuation on the boundary for Dini domains.* Proc. Amer. Math. Soc. **126** (1998), 441–446.

[34] G.M. Lieberman, *Regularized distance and its applications.* Pacific J. Math. **117** (1985), 329–353.

[35] L. Liu, *Stability estimates for the two dimensional inverse conductivity problem*, Ph.D. thesis, Department of Mathematics, University of Rochester, New York, 1997.

[36] A. Morassi and E. Rosset, *Stable determination of cavities in elastic bodies.* Inverse Problems **20** (2004), 453–480.

[37] A.I. Nachman, *Global uniqueness for a two-dimensional inverse boundary value problem.* Ann. of Math. **142** (1995), 71–96.

[38] G. Nakamura, *Inverse problems for elasticity.* Amer. Math. Soc. Transl. Ser. 2 **211** (2003), 71–85.

[39] G. Nakamura and G. Uhlmann, *Identification of Lamé parameters by boundary measurements.* Amer. J. Math. **115** (1993), 1161–1187.

[40] G. Nakamura and G. Uhlmann, *Global uniqueness for an inverse boundary problem arising in elasticity.* Invent. Math. **118** (1994), 457–474.

[41] G. Nakamura and G. Uhlmann, *Inverse problem at the boundary for an elastic medium.* SIAM J. Math. Anal. **26** (1995), 263–279.

[42] G. Nakamura and G. Uhlmann, *Erratum: Global uniqueness for an inverse boundary value problem arising in elasticity.* Invent. Math. **152** (2003), 205–207.

[43] J. Sylvester and G. Uhlmann, *A global uniqueness theorem for an inverse boundary value problem.* Ann. of Math. **125** (1987), 153–169.

[44] G. Uhlmann, *Developments in inverse problems since Calderón's foundational paper*, in *Harmonic Analysis and Partial Differential Equations*, Essays in Honor of Alberto P. Calderón, Chicago Lectures in Math., Univ. Chicago Press, Chicago, 1999, pp. 295–345.

[45] N. Weck, *Außenraumaufgaben in der Theorie stationärer Schwingungen inhomogener elasticher Körper.* Math. Z. **111** (1969), 387–398.

Editorial Information

To be published in the *Memoirs*, a paper must be correct, new, nontrivial, and significant. Further, it must be well written and of interest to a substantial number of mathematicians. Piecemeal results, such as an inconclusive step toward an unproved major theorem or a minor variation on a known result, are in general not acceptable for publication.

Papers appearing in *Memoirs* are generally at least 80 and not more than 200 published pages in length. Papers less than 80 or more than 200 published pages require the approval of the Managing Editor of the Transactions/Memoirs Editorial Board.

As of March 31, 2009, the backlog for this journal was approximately 12 volumes. This estimate is the result of dividing the number of manuscripts for this journal in the Providence office that have not yet gone to the printer on the above date by the average number of monographs per volume over the previous twelve months, reduced by the number of volumes published in four months (the time necessary for preparing a volume for the printer). (There are 6 volumes per year, each usually containing at least 4 numbers.)

A Consent to Publish and Copyright Agreement is required before a paper will be published in the *Memoirs*. After a paper is accepted for publication, the Providence office will send a Consent to Publish and Copyright Agreement to all authors of the paper. By submitting a paper to the *Memoirs*, authors certify that the results have not been submitted to nor are they under consideration for publication by another journal, conference proceedings, or similar publication.

Information for Authors

Memoirs are printed from camera copy fully prepared by the author. This means that the finished book will look exactly like the copy submitted.

Initial submission. The AMS uses Centralized Manuscript Processing for initial submissions. Authors should submit a PDF file using the Initial Manuscript Submission form found at `www.ams.org/peer-review-submission`, or send one copy of the manuscript to the following address: Centralized Manuscript Processing, MEMOIRS OF THE AMS, 201 Charles Street, Providence, RI 02904-2294 USA. If a paper copy is being forwarded to the AMS, indicate that it is for it Memoirs and include the name of the corresponding author, contact information such as email address or mailing address, and the name of an appropriate Editor to review the paper (see the list of Editors below).

The paper must contain a *descriptive title* and an *abstract* that summarizes the article in language suitable for workers in the general field (algebra, analysis, etc.). The *descriptive title* should be short, but informative; useless or vague phrases such as "some remarks about" or "concerning" should be avoided. The *abstract* should be at least one complete sentence, and at most 300 words. Included with the footnotes to the paper should be the 2000 *Mathematics Subject Classification* representing the primary and secondary subjects of the article. The classifications are accessible from `www.ams.org/msc/`. The list of classifications is also available in print starting with the 1999 annual index of *Mathematical Reviews*. The Mathematics Subject Classification footnote may be followed by a list of *key words and phrases* describing the subject matter of the article and taken from it. Journal abbreviations used in bibliographies are listed in the latest *Mathematical Reviews* annual index. The series abbreviations are also accessible from `www.ams.org/msnhtml/serials.pdf`. To help in preparing and verifying references, the AMS offers MR Lookup, a Reference Tool for Linking, at `www.ams.org/mrlookup/`.

Electronically prepared manuscripts. The AMS encourages electronically prepared manuscripts, with a strong preference for $\mathcal{A}_{\mathcal{M}}\mathcal{S}$-LaTeX. To this end, the Society has prepared $\mathcal{A}_{\mathcal{M}}\mathcal{S}$-LaTeX author packages for each AMS publication. Author packages include instructions for preparing electronic manuscripts, samples, and a style file that generates the particular design specifications of that publication series. Though $\mathcal{A}_\mathcal{M}\mathcal{S}$-LaTeX is the highly preferred format of TeX, author packages are also available in $\mathcal{A}_\mathcal{M}\mathcal{S}$-TeX.

Authors may retrieve an author package for *Memoirs of the AMS* from `www.ams.org/journals/memo/memoauthorpac.html` or via FTP to `ftp.ams.org` (login as `anonymous`, enter username as password, and type `cd pub/author-info`). The *AMS Author Handbook* and the *Instruction Manual* are available in PDF format from the author package link. The author package can also be obtained free of charge by sending email to `tech-support@ams.org` (Internet) or from the Publication Division, American Mathematical Society, 201 Charles St., Providence, RI 02904-2294, USA. When requesting an author package, please specify $\mathcal{A}_\mathcal{M}\mathcal{S}$-LaTeX or $\mathcal{A}_\mathcal{M}\mathcal{S}$-TeX and the publication in which your paper will appear. Please be sure to include your complete mailing address.

After acceptance. The final version of the electronic file should be sent to the Providence office (this includes any TeX source file, any graphics files, and the DVI or PostScript file) immediately after the paper has been accepted for publication.

Before sending the source file, be sure you have proofread your paper carefully. The files you send must be the EXACT files used to generate the proof copy that was accepted for publication. For all publications, authors are required to send a printed copy of their paper, which exactly matches the copy approved for publication, along with any graphics that will appear in the paper.

Accepted electronically prepared files can be submitted via the web at `www.ams.org/submit-book-journal/`, sent via FTP, or sent on CD-Rom or diskette to the Electronic Prepress Department, American Mathematical Society, 201 Charles Street, Providence, RI 02904-2294 USA. TeX source files, DVI files, and PostScript files can be transferred over the Internet by FTP to the Internet node `ftp.ams.org` (130.44.1.100). When sending a manuscript electronically via CD-Rom or diskette, please be sure to include a message identifying the paper as a Memoir.

Electronically prepared manuscripts can also be sent via email to `pub-submit@ams.org` (Internet). In order to send files via email, they must be encoded properly. (DVI files are binary and PostScript files tend to be very large.)

Electronic graphics. Comprehensive instructions on preparing graphics are available at `www.ams.org/authors/journals.html`. A few of the major requirements are given here.

Submit files for graphics as EPS (Encapsulated PostScript) files. This includes graphics originated via a graphics application as well as scanned photographs or other computer-generated images. If this is not possible, TIFF files are acceptable as long as they can be opened in Adobe Photoshop or Illustrator. No matter what method was used to produce the graphic, it is necessary to provide a paper copy to the AMS.

Authors using graphics packages for the creation of electronic art should also avoid the use of any lines thinner than 0.5 points in width. Many graphics packages allow the user to specify a "hairline" for a very thin line. Hairlines often look acceptable when proofed on a typical laser printer. However, when produced on a high-resolution laser imagesetter, hairlines become nearly invisible and will be lost entirely in the final printing process.

Screens should be set to values between 15% and 85%. Screens which fall outside of this range are too light or too dark to print correctly. Variations of screens within a graphic should be no less than 10%.

Inquiries. Any inquiries concerning a paper that has been accepted for publication should be sent to `memo-query@ams.org` or directly to the Electronic Prepress Department, American Mathematical Society, 201 Charles St., Providence, RI 02904-2294 USA.

Editors

This journal is designed particularly for long research papers, normally at least 80 pages in length, and groups of cognate papers in pure and applied mathematics. Papers intended for publication in the *Memoirs* should be addressed to one of the following editors. The AMS uses Centralized Manuscript Processing for initial submissions to AMS journals. Authors should follow instructions listed on the Initial Submission page found at www.ams.org/memo/memosubmit.html.

Algebra to ALEXANDER KLESHCHEV, Department of Mathematics, University of Oregon, Eugene, OR 97403-1222; email: ams@noether.uoregon.edu

Algebraic geometry to DAN ABRAMOVICH, Department of Mathematics, Brown University, Box 1917, Providence, RI 02912; email: amsedit@math.brown.edu

Algebraic geometry and its applications to MINA TEICHER, Emmy Noether Research Institute for Mathematics, Bar-Ilan University, Ramat-Gan 52900, Israel; email: teicher@macs.biu.ac.il

Algebraic topology to ALEJANDRO ADEM, Department of Mathematics, University of British Columbia, Room 121, 1984 Mathematics Road, Vancouver, British Columbia, Canada V6T 1Z2; email: adem@math.ubc.ca

Combinatorics to JOHN R. STEMBRIDGE, Department of Mathematics, University of Michigan, Ann Arbor, Michigan 48109-1109; email: JRS@umich.edu

Commutative and homological algebra to LUCHEZAR L. AVRAMOV, Department of Mathematics, University of Nebraska, Lincoln, NE 68588-0130; email: avramov@math.unl.edu

Complex analysis and harmonic analysis to ALEXANDER NAGEL, Department of Mathematics, University of Wisconsin, 480 Lincoln Drive, Madison, WI 53706-1313; email: nagel@math.wisc.edu

Differential geometry and global analysis to CHRIS WOODWARD, Department of Mathematics, Rutgers University, 110 Frelinghuysen Road, Piscataway, NJ 08854; email: ctw@math.rutgers.edu

Dynamical systems and ergodic theory and complex analysis to YUNPING JIANG, Department of Mathematics, CUNY Queens College and Graduate Center, 65-30 Kissena Blvd., Flushing, NY 11367; email: Yunping.Jiang@qc.cuny.edu

Functional analysis and operator algebras to DIMITRI SHLYAKHTENKO, Department of Mathematics, University of California, Los Angeles, CA 90095; email: shlyakht@math.ucla.edu

Geometric analysis to WILLIAM P. MINICOZZI II, Department of Mathematics, Johns Hopkins University, 3400 N. Charles St., Baltimore, MD 21218; email: trans@math.jhu.edu

Geometric topology to MARK FEIGHN, Math Department, Rutgers University, Newark, NJ 07102; email: feighn@andromeda.rutgers.edu

Harmonic analysis, representation theory, and Lie theory to ROBERT J. STANTON, Department of Mathematics, The Ohio State University, 231 West 18th Avenue, Columbus, OH 43210-1174; email: stanton@math.ohio-state.edu

Logic to STEFFEN LEMPP, Department of Mathematics, University of Wisconsin, 480 Lincoln Drive, Madison, Wisconsin 53706-1388; email: lempp@math.wisc.edu

Number theory to JONATHAN ROGAWSKI, Department of Mathematics, University of California, Los Angeles, CA 90095; email: jonr@math.ucla.edu

Number theory to SHANKAR SEN, Department of Mathematics, 505 Malott Hall, Cornell University, Ithaca, NY 14853; email: ss70@cornell.edu

Partial differential equations to GUSTAVO PONCE, Department of Mathematics, South Hall, Room 6607, University of California, Santa Barbara, CA 93106; email: ponce@math.ucsb.edu

Partial differential equations and dynamical systems to PETER POLACIK, School of Mathematics, University of Minnesota, Minneapolis, MN 55455; email: polacik@math.umn.edu

Probability and statistics to RICHARD BASS, Department of Mathematics, University of Connecticut, Storrs, CT 06269-3009; email: bass@math.uconn.edu

Real analysis and partial differential equations to DANIEL TATARU, Department of Mathematics, University of California, Berkeley, Berkeley, CA 94720; email: tataru@math.berkeley.edu

All other communications to the editors should be addressed to the Managing Editor, ROBERT GURALNICK, Department of Mathematics, University of Southern California, Los Angeles, CA 90089-1113; email: guralnic@math.usc.edu.

Titles in This Series

941 **Gelu Popescu,** Unitary invariants in multivariable operator theory, 2009

940 **Gérard Iooss and Pavel I. Plotnikov,** Small divisor problem in the theory of three-dimensional water gravity waves, 2009

939 **I. D. Suprunenko,** The minimal polynomials of unipotent elements in irreducible representations of the classical groups in odd characteristic, 2009

938 **Antonino Morassi and Edi Rosset,** Uniqueness and stability in determining a rigid inclusion in an elastic body, 2009

937 **Skip Garibaldi,** Cohomological invariants: Exceptional groups and spin groups, 2009

936 **André Martinez and Vania Sordoni,** Twisted pseudodifferential calculus and application to the quantum evolution of molecules, 2009

935 **Mihai Ciucu,** The scaling limit of the correlation of holes on the triangular lattice with periodic boundary conditions, 2009

934 **Arjen Doelman, Björn Sandstede, Arnd Scheel, and Guido Schneider,** The dynamics of modulated wave trains, 2009

933 **Luchezar Stoyanov,** Scattering resonances for several small convex bodies and the Lax-Phillips conjuecture, 2009

932 **Jun Kigami,** Volume doubling measures and heat kernel estimates of self-similar sets, 2009

931 **Robert C. Dalang and Marta Sanz-Solé,** Hölder-Sobolv regularity of the solution to the stochastic wave equation in dimension three, 2009

930 **Volkmar Liebscher,** Random sets and invariants for (type II) continuous tensor product systems of Hilbert spaces, 2009

929 **Richard F. Bass, Xia Chen, and Jay Rosen,** Moderate deviations for the range of planar random walks, 2009

928 **Ulrich Bunke,** Index theory, eta forms, and Deligne cohomology, 2009

927 **N. Chernov and D. Dolgopyat,** Brownian Brownian motion-I, 2009

926 **Riccardo Benedetti and Francesco Bonsante,** Canonical wick rotations in 3-dimensional gravity, 2009

925 **Sergey Zelik and Alexander Mielke,** Multi-pulse evolution and space-time chaos in dissipative systems, 2009

924 **Pierre-Emmanuel Caprace,** "Abstract" homomorphisms of split Kac-Moody groups, 2009

923 **Michael Jöllenbeck and Volkmar Welker,** Minimal resolutions via algebraic discrete Morse theory, 2009

922 **Ph. Barbe and W. P. McCormick,** Asymptotic expansions for infinite weighted convolutions of heavy tail distributions and applications, 2009

921 **Thomas Lehmkuhl,** Compactification of the Drinfeld modular surfaces, 2009

920 **Georgia Benkart, Thomas Gregory, and Alexander Premet,** The recognition theorem for graded Lie algebras in prime characteristic, 2009

919 **Roelof W. Bruggeman and Roberto J. Miatello,** Sum formula for SL_2 over a totally real number field, 2009

918 **Jonathan Brundan and Alexander Kleshchev,** Representations of shifted Yangians and finite W-algebras, 2008

917 **Salah-Eldin A. Mohammed, Tusheng Zhang, and Huaizhong Zhao,** The stable manifold theorem for semilinear stochastic evolution equations and stochastic partial differential equations, 2008

916 **Yoshikata Kida,** The mapping class group from the viewpoint of measure equivalence theory, 2008

For a complete list of titles in this series, visit the
AMS Bookstore at **www.ams.org/bookstore/**.